CHALLENGES FOR THE CHEMICAL SCIENCES IN THE 21ST CENTURY

THE
ENVIRONMENT

ORGANIZING COMMITTEE FOR THE WORKSHOP ON
THE ENVIRONMENT

COMMITTEE ON CHALLENGES FOR THE CHEMICAL SCIENCES
IN THE 21ST CENTURY

BOARD ON CHEMICAL SCIENCES AND TECHNOLOGY

DIVISION ON EARTH AND LIFE STUDIES

NATIONAL RESEARCH COUNCIL
OF THE NATIONAL ACADEMIES

THE NATIONAL ACADEMIES PRESS
Washington, D.C.
www.nap.edu

NOTICE: The project that is the subject of this report was approved by the Governing Board of the National Research Council, whose members are drawn from the councils of the National Academy of Sciences, the National Academy of Engineering, and the Institute of Medicine. The members of the committee responsible for the report were chosen for their special competences and with regard for appropriate balance.

Support for this study was provided by the National Research Council, the U.S. Department of Energy (DE-AT-010EE41424, BES DE-FG-02-00ER15040, and DE-AT01-03ER15386), the National Science Foundation (CTS-9908440), the Defense Advanced Research Projects Agency (DOD MDA972-01-M-0001), the U.S. Environmental Protection Agency (R82823301 and X83080801), the American Chemical Society, the American Institute of Chemical Engineers, the Camille and Henry Dreyfus Foundation, Inc. (SG00-093), the National Institute of Standards and Technology (NA1341-01-2-1070 and 43NANB010995), the National Institutes of Health (NCI-N01-OD-4-2139 and NIGMS-N01-OD-4-2139), the Green Chemistry Institute, and the chemical industry

All opinions, findings, conclusions, or recommendations expressed herein are those of the authors and do not necessarily reflect the views of the organizations or agencies that provided support for this project.

THE NATIONAL ACADEMIES
Advisers to the Nation on Science, Engineering, and Medicine

The **National Academy of Sciences** is a private, nonprofit, self-perpetuating society of distinguished scholars engaged in scientific and engineering research, dedicated to the furtherance of science and technology and to their use for the general welfare. Upon the authority of the charter granted to it by the Congress in 1863, the Academy has a mandate that requires it to advise the federal government on scientific and technical matters. Dr. Bruce M. Alberts is president of the National Academy of Sciences.

The **National Academy of Engineering** was established in 1964, under the charter of the National Academy of Sciences, as a parallel organization of outstanding engineers. It is autonomous in its administration and in the selection of its members, sharing with the National Academy of Sciences the responsibility for advising the federal government. The National Academy of Engineering also sponsors engineering programs aimed at meeting national needs, encourages education and research, and recognizes the superior achievements of engineers. Dr. Wm. A. Wulf is president of the National Academy of Engineering.

The **Institute of Medicine** was established in 1970 by the National Academy of Sciences to secure the services of eminent members of appropriate professions in the examination of policy matters pertaining to the health of the public. The Institute acts under the responsibility given to the National Academy of Sciences by its congressional charter to be an adviser to the federal government and, upon its own initiative, to identify issues of medical care, research, and education. Dr. Harvey V. Fineberg is president of the Institute of Medicine.

The **National Research Council** was organized by the National Academy of Sciences in 1916 to associate the broad community of science and technology with the Academy's purposes of furthering knowledge and advising the federal government. Functioning in accordance with general policies determined by the Academy, the Council has become the principal operating agency of both the National Academy of Sciences and the National Academy of Engineering in providing services to the government, the public, and the scientific and engineering communities. The Council is administered jointly by both Academies and the Institute of Medicine. Dr. Bruce M. Alberts and Dr. Wm. A. Wulf are chair and vice chair, respectively, of the National Research Council.

www.national-academies.org

ORGANIZING COMMITTEE FOR THE
WORKSHOP ON THE ENVIRONMENT

MARIO J. MOLINA, Massachusetts Institute of Technology, *Co-Chair*
JOHN H. SEINFELD, California Institute of Technology, *Co-Chair*
PHILIP H. BRODSKY, Pharmacia (retired)
JEAN H. FUTRELL, Pacific Northwest National Laboratory
PARRY M. NORLING, RAND Corporation
CHRISTINE S. SLOANE, General Motors Corporation
ISIAH M. WARNER, Louisiana State University

Liaisons
MARK A. BARTEAU, University of Delaware (Steering Committee Liaison)
A. WELFORD CASTLEMAN, JR., The Pennsylvania State University (BCST
    Liaison)
JOSEPH M. DeSIMONE, University of North Carolina, Chapel Hill, and North
    Carolina State University (BCST Liaison)
JEFFREY J. SIIROLA, Eastman Chemical Company (Steering Committee Liaison)
ROBERT M. SUSSMAN, Latham & Watkins (BCST Liaison)

Staff

JENNIFER J. JACKIW, Program Officer
SYBIL A. PAIGE, Administrative Associate
DOUGLAS J. RABER, Senior Scholar
DAVID C. RASMUSSEN, Program Assistant

COMMITTEE ON CHALLENGES FOR THE CHEMICAL SCIENCES IN THE 21ST CENTURY

RONALD BRESLOW, Columbia University, *Co-Chair*
MATTHEW V. TIRRELL, University of California at Santa Barbara, *Co-Chair*
MARK A. BARTEAU, University of Delaware
JACQUELINE K. BARTON, California Institute of Technology
CAROLYN R. BERTOZZI, University of California at Berkeley
ROBERT A. BROWN, Massachusetts Institute of Technology
ALICE P. GAST,[1] Stanford University
IGNACIO E. GROSSMANN, Carnegie Mellon University
JAMES M. MEYER,[2] DuPont Co.
ROYCE W. MURRAY, University of North Carolina at Chapel Hill
PAUL J. REIDER, Amgen, Inc.
WILLIAM R. ROUSH, University of Michigan
MICHAEL L. SHULER, Cornell University
JEFFREY J. SIIROLA, Eastman Chemical Company
GEORGE M. WHITESIDES, Harvard University
PETER G. WOLYNES, University of California, San Diego
RICHARD N. ZARE, Stanford University

Staff

JENNIFER J. JACKIW, Program Officer
CHRISTOPHER K. MURPHY, Program Officer
SYBIL A. PAIGE, Administrative Associate
DOUGLAS J. RABER, Senior Scholar
DAVID C. RASMUSSEN, Program Assistant

---

[1]Committee member until July 2001; subsequently the Board on Chemical Sciences and Technology (BCST) liaison to the committee in her role as BCST co-chair.

[2]Committee member until March 2002, following his retirement from DuPont.

Preface

The Workshop on the Environment was held in Irvine, California, on November 17-19, 2002. This workshop was the third in a series of six workshops that make up the study Challenges for the Chemical Sciences in the 21st Century. The task for each of the workshops was defined as follows:

Each workshop—and its subsequent report—will address a series of common themes:

• Discovery: Identify major discoveries or advances in the chemical sciences during the last several decades.
• Interfaces: Identify the major discoveries and challenges at the interfaces between chemistry/chemical engineering and such areas as biology, environmental science, materials science, medicine, and physics.
• Challenges: Identify the grand challenges that exist in the chemical sciences.
• Infrastructure: Identify the issues and opportunities that exist in the chemical sciences to improve the infrastructure for research and education, and demonstrate the value of these activities to society.

The Workshop on the Environment brought together a diverse group of participants (Appendix F) from the chemical sciences who were addressed by invited speakers in plenary session on a variety of issues and challenges for the chemical sciences as they relate to environmental science and technology (Appendix C). These presentations served as a starting point for discussions and comments by the participants. The workshop participants were then divided into small groups that met periodically during the workshop to further discuss and analyze the relevant issues. Each group reported its discussions to the workshop as a whole.

This report is intended to reflect the concepts discussed and opinions expressed at the Workshop on the Environment and is not intended to be a comprehensive overview of all of the potential challenges that exist for the chemical sciences in the areas of environmental science and technology. The organizing committee has used this input from workshop participants as a basis for the conclusions expressed in this report. However, sole responsibility for these conclusions rests with the organizing committee.

Although much progress has been achieved toward detailed mechanistic understanding in all environmental sciences, the greatest advances appear to have occurred in the atmospheric chemistry field, reflecting most likely the greater complexity of aquatic and terrestrial systems and their biological components.

This study was conducted under the auspices of the National Research Council's Board on Chemical Sciences and Technology, with assistance provided by its staff. The committee acknowledges this support.

A number of our colleagues across the chemical sciences community provided the committee with helpful input. The committee thanks Patrick G. Hatcher, Robert G. Keesee, Karl T. Mueller, and Paul Ziemann.

Mario J. Molina and John H. Seinfeld, Co-Chairs,
Organizing Committee for the Workshop on the Environment
Challenges for the Chemical Sciences in the 21st Century

Acknowledgment of Reviewers

This report has been reviewed in draft form by individuals chosen for their diverse perspectives and technical expertise, in accordance with procedures approved by the National Research Council's (NRC's) Report Review Committee. The purpose of this independent review is to provide candid and critical comments that will assist the institution in making the published report as sound as possible and to ensure that the report meets institutional standards for objectivity, evidence, and responsiveness to the study charge. The review comments and draft manuscript remain confidential to protect the integrity of the deliberative process. We wish to thank the following individuals for their participation in the review of this report:

Heather C. Allen, The Ohio State University
Liese Dallbauman, Gas Technology Institute
Thom H. Dunning, Jr., Oak Ridge National Laboratory and University of
    Tennessee
William H. Glaze, Oregon Health & Science University
George M. Hornberger, University of Virginia
Henry T. Kohlbrand, The Dow Chemical Company
Stephen G. Maroldo, Rohm and Haas Company
Peggy O'Day, University of California, Merced
Ed Yeung, Iowa State University

Although the reviewers listed above have provided many constructive comments and suggestions, they were not asked to endorse the conclusions or recommendations nor did they see the final draft of the report before its release.

The review of this report was overseen by George E. Keller II (Union Carbide, retired). Appointed by the National Research Council, he was responsible for making certain that an independent examination of this report was carried out in accordance with institutional procedures and that all review comments were carefully considered. Responsibility for the final content of this report rests entirely with the authoring committee and the institution.

Contents

Executive Summary

The chemical sciences have made enormous contributions to the well-being of our nation and to that of the world—as an economic driver and through the manufactured substances that range from life-saving pharmaceuticals to materials that improve the safety of our automobiles. At the same time, the relationship between the chemical sciences and the environment is on less firm ground. Indeed, to a great extent the environmental movement in the 1970s began in response to harmful effects of certain industrial chemicals. It matters little that those harmful effects were unintentional, because the harm was real.

Over the last several decades, chemists and chemical engineers—indeed, the entire chemical enterprise—have made remarkable progress on several fronts. Earlier practices—of disposing of chemical waste directly into the air, waterways, and soil—largely have been replaced with approaches that have much lower environmental impact. Industrial and automotive smog have been greatly reduced, our rivers and streams are much cleaner than they were three decades ago, and we have almost completely eliminated the use of chlorofluorocarbons (CFCs) and other halogenated compounds that cause ozone depletion in the stratosphere. Modern pesticides are safer for the environment and are used in much smaller quantities, and the chemical industry has adopted a program of Responsible Care[1] in which it practices stewardship of the chemicals it produces from cradle to grave. Perhaps most importantly, the research activities of chemists and chemical engineers have made huge strides toward understanding the chemistry of the environment at a fundamental level.

---

[1] *http://www.americanchemistry.com/*

1

However, much remains to be done. The challenge is global in nature: not only do countries around the world face environmental problems, but pollutants generated in one location are transported across continents and oceans. Similarly, the different parts of the environment are interconnected, and attempting to manage environmental sinks separately will only move the problem from one phase to another. The magnitude and scope of the problem is huge, and the inescapability of environmental problems demands major effort in science and engineering. In November 2002, as part of Challenges for the Chemical Sciences in the 21st Century, the Board on Chemical Sciences and Technology convened the Workshop on the Environment in Irvine, California. The workshop organizing committee assembled a group of top environmental scientists to deliver plenary lectures (Appendix C), and an outstanding group of chemical scientists and engineers—from academia, government, national laboratories, and industrial laboratories—was recruited to participate in the workshop (Appendix F). Through the use of extensive discussion periods and breakout sessions, input from the entire group of participants was obtained during the course of the workshop. The results of the breakout sessions are presented in Appendix G, and written versions of the speakers' presentations are provided in Appendix D. In combination with other references cited in this report, the data collected at the workshop provide the basis for this report.

The structure of the Workshop on the Environment followed that of the parent project and each of the other workshops that were held as part of the study of Challenges for the Chemical Sciences in the 21st Century (Materials and Manufacturing, Energy and Transportation, National Security and Homeland Defense, Information and Communications, and Health and Medicine). Under this structure, the workshop addressed four specific themes:

1. *Discovery:* What major discoveries or advances related to the environment have been made in the chemical sciences during the last several decades?

2. *Interfaces:* What are the major environment-related discoveries and challenges at the interfaces between chemistry–chemical engineering and other disciplines, including biology, information science, materials science, and physics?

3. *Challenges:* What are the environment-related grand challenges in the chemical sciences and engineering?

4. *Infrastructure:* What are the issues at the intersection of environmental studies and the chemical sciences for which there are structural challenges and opportunities—in teaching, research, equipment, codes and software, facilities, and personnel?

Workshop participants provided a broad range of experience and perspective, and their discussions and presentations identified a wide variety of opportunities and challenges in chemistry and chemical engineering. These opportunities and challenges, which are documented throughout this report (including many of the specifics described in Appendix D and Appendix G), led the committee to its overarching conclusions.

Conclusion: Chemistry and chemical engineering have made major contributions to solving environmental problems.

Specific areas of accomplishment include

- major increases in analytical capabilities—detection, monitoring, and measurement;
- increased understanding of biogeochemical processes and cycles;
- advances in industrial ecology—new attitudes about pollution prevention;
- development of environmentally benign materials (e.g., CFC replacements);
- new methods for waste treatment and pollution prevention;
- green chemistry and new chemical processes;
- discovery of environmental problems and identification of their underlying causes and mechanisms; and
- development of improved modeling and simulation techniques.

Conclusion: Collaboration of chemists and chemical engineers with scientists and engineers in other disciplines has led to important discoveries.

These contributions have enhanced both basic understanding and the solution of environmental problems through work at the interfaces of the chemical sciences with biology, physics, engineering, materials science, mathematics, computer science, atmospheric science, meteorology, and geology.

Conclusion: Manifold challenges and opportunities in chemistry and chemical engineering exist at the interface with the environmental sciences.

By responding to these opportunities and challenges, the chemical sciences community will be able to make substantial contributions to

- fundamental understanding of the environment,
- remediation of environmental problems that currently exist,
- prevention of environmental problems in the future, and
- protection of the environment.

The stakes for responding to these challenges are high because regulatory decisions might cost or preserve billions of dollars, impact millions of human lives, or even determine the fate of entire species.

Much of the discussion at the workshop emphasized the interrelated nature of the many parts of the environment. Typically, it is not possible to take action in one area without creating at least the possibility of impacting other areas as well. In order to avoid such undesired consequences, a systems approach will be needed for the discovery and management of problems of the atmosphere, water, and soil. This will be necessary not only for understanding the complexity of each medium but for avoiding regulatory-driven tendencies to simply shift impacts from one medium to another.

A life-cycle systems approach, similar to what has been developed to evaluate energy impacts, will facilitate sound management of environmental impacts. This will provide a clear understanding of both where and when environmental impacts occur in the life of a product, process, or service. It also will make it possible to appreciate all impacts, and to see how interactions and alternatives at each point in a life cycle can influence other parts of the life cycle. For chemical processing and manufacturing, significant impact can occur at various stages, including extraction and preparation of raw materials, conversion of raw materials into products, separation and purification of materials, product distribution, end use of products, and final disposition after the useful life of products.

Conclusion: A systems approach is essential for solution of environmental problems.

The systems approach will be needed in several areas, including

• actions that affect any of the three principal environmental sinks (air, water, and soil) and the biological systems with which they interact, where attempts to manage each of them separately will surely transfer impacts from one medium to another;

• spatial management of environmental-impact sources—where the impacts are generated in a processing and manufacturing sequence; and

• temporal management of environmental-impact sources—when the impacts are generated in a processing and manufacturing sequence.

The use of systems approaches will necessitate simulation and modeling of enormously large and complex systems. This will require significant computational resources, intensive efforts in complex optimization, and formulation of mathematical models. Input and expertise from a broad array of scientific, engineering, and social disciplines will be an essential part of developing the necessary tools.

Conclusion: Solving environmental problems will require intensive mathematical modeling, complex optimization, and computational resources.

Systems approaches will necessitate extensive collaborations among a wide range of scientific, engineering, and social disciplines. Modeling of large environmental systems (e.g., climate modeling that spans large temporal and spatial ranges) will generate massive datasets, fast and robust networks, and powerful computers for processing the data arrays.

A broad array of research challenges face the chemical sciences, both in areas of fundamental understanding and for specific environmental problems.

Conclusion: Important opportunities exist for chemists and chemical engineers to contribute to a better understanding of the environment.

Many of these research opportunities will involve work at the interfaces with other disciplines or interdisciplinary collaborations with scientists and engineers from those disciplines. Just as these collaborations have led to significant progress in the past, they should be expected to play an important role in the future to fully understand and solve environmental problems. Examples include the need to understand (or better understand)

- structure-toxicity relations;
- chemical processes at the molecular level;
- biological and physicochemical interactions in response to environmental stresses;
- fate and transport of anthropogenic chemicals;
- biogeochemical cycles;
- gas-to-particle conversion in the atmosphere;
- functional genomics and the chemical processes that govern organism-environment relationships; and
- chemical-gene interactions in the real environment.

As we continue to better understand the underlying science of the environment, further advances will require new tools and instruments.

Conclusion: Chemists and chemical engineers will need to develop new analytical instruments and tools.

These tools and instruments will have to function effectively in an increasingly complex research arena that involves measurements of vanishingly small quantities of substances in the presence of contamination from other chemicals, under circumstances that make sample acquisition difficult. They will have to address three principal areas of measurement:

1. laboratory analyses
2. field measurements
3. theoretical tools for modeling and comparison with experiment

Conclusion: Improved methods for sampling and monitoring must be developed.

Chemists and chemical engineers will have to address the challenges of sampling and monitoring—air, water, and soil—more extensively and more frequently than can be done now. This will require improvements in instrumentation, in sampling methodology, and in techniques for remote measurements.

Conclusion: The new approaches of green chemistry and sustainable chemistry offer the potential for developing chemical and manufacturing processes that are environmentally beneficial.

We are still in the early stages, but successful examples already have been reported. If the necessary investment is made in these new directions, chemists and chemical engineers will be able to make major strides in improving environmental quality.

Conclusion: Strong and continued support of the chemical sciences will be an essential part of the federal research investment for understanding, improving, and protecting the environment.

Chemists and chemical engineers will be able to respond effectively to the challenges described here only if they have the resources needed to carry out the necessary research. This impact of support will be enhanced if it facilitates interdisciplinary research and encourages industrial partnerships. The scientific progress resulting from such support will inform and enable the policy-making and decision process that is essential to future environmental improvement.

1

Introduction

For millennia, advances in human progress have been tied to our ability to protect ourselves from the harmful effects of the wastes we produce—ranging from human waste to the organic and inorganic by-products of everyday living. Across the world, cultures learned to bury their dead away from their homes and to burn their waste or make certain that it was carried away by streams and rivers flowing downstream from their homes. Those cultures that learned this most effectively thrived. When the industrial revolution took place in the nineteenth century, rivers again enabled progress. They provided water needed for power and energy, and they carried away the waste materials from industrial processes.

However, things had changed by the middle of the twentieth century. The increase in human population and the growth of modern industry were leading to signs that the system was overloaded. There were reports of rivers that had turned orange or had caught on fire, the smog over some cities was becoming intolerable, and there were signs of negative health consequences from buried waste. The methods of waste disposal that had helped us build our modern society were turning back on us. Public attention was captured by Rachel Carson's book *Silent Spring*,[1] and events surrounding the Vietnam war were changing the political landscape. In 1970 the U.S. Environmental Protection Agency was formed, and the Clean Air Act was passed later that same year. Additional legislation followed to cover other areas of the environment.

Many of the problems that needed to be solved were chemical in nature, and the chemical industry was seen by many as the source of our environmental prob-

---

[1]Rachel Carson, *Silent Spring;* Houghton Mifflin: New York, 1962.

lems. Suddenly we had moved beyond industry and modern technology as the source of our high quality of life. The chemical industry was no longer viewed in a positive light. As with most such situations, there were conflicts between regulatory policy and the financial interests of the companies being regulated, and progress was sometimes slow. But changes have been made. The rivers are cleaner, and the smog has decreased. Bird populations are no longer suffering the effects of DDT, and disposal of chemical waste is carried out in a safer and more reliable manner. The chemical industry, through the American Chemistry Council, has established a strong industry standard with its *Responsible Care* program.

The committee organized a workshop that was held in Irvine, California, in November 2002, to address ways in which chemists and chemical engineers could focus their R&D efforts on the solution of environmental problems. This report is part of a broader project, Challenges for the Chemical Sciences in the 21st Century. The overview report for the project includes a chapter on Atmospheric and Environmental Chemistry.[2] A series of speakers (Appendix F) presented lectures (Appendix D) on topics that covered all parts of the environment—the biosphere, the atmosphere, soil, and water. They addressed issues in manufacturing, energy production, and remediation of those parts of the environment that already have suffered damage. Considerable input for the report was also provided by a series of breakout sessions (Appendix G) in which all workshop attendees participated (Appendix E). These breakout sessions explored the ways in which chemists and chemical engineers already have contributed to solving environmental problems, the technical challenges that they can help to overcome in the future, and the barriers that will have to be overcome for them to do so. The specific questions addressed in the four breakout sessions were the following:

- **Discovery:** What major discoveries or advances related to the environment have been made in the chemical sciences during the last several decades?
- **Interfaces:** What are the major environment-related discoveries and challenges at the interfaces between chemistry–chemical engineering and other disciplines, including biology, information science, materials science, and physics?
- **Challenges:** What are the environment-related grand challenges in the chemical sciences and engineering?
- **Infastructure:** What are the issues at the intersection of environmental studies and the chemical sciences for which there are structural challenges and opportunities—in teaching, research, equipment and instrumentation, facilities, and personnel?

---

[2]*Beyond the Molecular Frontier: Challenges for Chemistry and Chemical Engineering,* National Research Council, The National Academies Press, Washington, D.C., 2003.

We've seen much progress in the past few decades, but more remains to be done. Some regulations are in place, while others are still being developed. The anxiety over global climate change has introduced an entirely new set of concerns in the last decade, with conflicting proposals about how the world should respond. One thing is certain, however, and that is the need for the chemical sciences community to participate in solving the problems.

2

Successes and Discoveries in Environmental Chemical Science

Progress in environmental science during the last several decades has depended heavily on contributions from chemistry and chemical engineering. These contributions range from understanding fundamental concepts of the behavior of materials in the environment to the development of procedures for protecting the environment and remediating environmentally contaminated sites. In the areas of fundamental science and engineering, the contributions fall into the categories of new or enhanced analytical capabilities, advances in fundamental science, and the development of new models and databases. Chemical contributions in the areas of environmental protection and remediation can be classified into technologies for pollution control and remediation and for pollution prevention via synthesis, manufacturing, and process advances.

ANALYTICAL CAPABILITIES

Developments in analytical capabilities have allowed chemists and chemical engineers to detect ever smaller quantities of chemical substances. One hundred years ago, chemists were challenged by the task of carrying out analysis on samples smaller than a gram. But in some cases it is now possible to detect and measure the presence of substances at the level of single molecules, an improvement of some 20 orders of magnitude. Increases in time resolution have been similarly astonishing, from a situation in which it was difficult to measure events taking place at a time scale of less than a second to the investigation of processes that take place on the femtosecond (10^{-15} s) time scale. Often the measurements must be made remotely or must examine substances that are present in vanishingly small concentrations.

Some of the improvements in detection, monitoring, and measurement science that have contributed to advances in environmental science include the development of the electron-capture detector (with its high sensitivity for halogenated organic compounds), the application of gas chromatography-mass spectrometry (which combines separation of mixtures with reliable identification of their components), advances in sensor technology, and the availability of synchrotron-based methods for x-ray studies. Improved mass spectrometric methods have made it possible to characterize single aerosol particles in the atmosphere and have provided improved time resolution for atmospheric measurements. Satellite-based technology and remote sensing have enabled monitoring and space-based measurements that are far more comprehensive than those that were possible using aircraft and balloons.

FUNDAMENTAL CHEMICAL SCIENCE

Advances in the understanding of basic chemical science have been responsible for substantial progress in environmental science. Similarly, progress in environmental chemistry has driven fundamental science, as illustrated by the vastly improved understanding of the nature of complex bimolecular reactions brought about by investigations of important atmospheric chemical processes. Key contributions include the following:

- advances in fundamental understanding of the chemistry of free radicals, which play key roles in a variety of atmospheric, aquatic, and terrestrial chemical processes;
- advances in surface chemistry that have provided a better understanding of reactions on surfaces and in microporous regions;
- better understanding of homogeneous gas-phase chemistry;
- establishing structure-activity relationships for the activity and fate of chemical species in the environment;
- advances in genetics at the molecular level that have made it possible to use bioremediation to clean up environmentally contaminated sites and to understand generic gene-environment relationships in the environment;
- fundamental understanding of the internal combustion process, which has led to improved engine efficiency; and
- advances in chemically based technologies (e.g., fuel cells, batteries, and photovoltaics) that have contributed to environmental improvements in the energy and transportation sectors.[1]

---

[1] Also see the related report in this series: *Challenges for the Chemical Sciences in the 21st Century: Energy and Transportation,* National Research Council, The National Academies Press, Washington, D.C., 2003.

Some of the advances in fundamental science that have led to new attitudes and approaches to environmental problems include the use of correlated chemical measurements to investigate environmental processes, development of industrial ecology as a framework for studying issues, use of life-cycle analysis to evaluate the impact of substances in the environment, recognition of the importance of speciation (in contrast to total concentration) of chemical substances, and recognition that a systems approach is often necessary to address complex environmental issues.

MODELS AND DATABASES

By any criteria one might employ, the environment is a large and complex system. As a consequence, there are large barriers to developing predictive capabilities, even for localized portions of the environment. However, these barriers have been partially surmounted by the substantial progress that has been made in computing-related areas, enabled in large part by the tremendous growth over the last several decades in computing speed and capabilities. Specific areas of accomplishment include the generation of databases of kinetic and thermodynamic data, new software and modeling techniques (at scales from molecular to global), development of computational models that permit the simulation of both fundamental and complex systems with ever increasing fidelity, and to some extent, advances in predicting risk and taking risk-based corrective action.

POLLUTION CONTROL, REMEDIATION, AND PURIFICATION

Technical solutions have been developed to a number of important environmental problems. One of the most important contributions to human health has been the chemical purification of drinking water, which has nearly eliminated water-borne diseases in developed countries.[2] Disinfection with chlorine and ozonation have been used to eliminate pathogens, and advances in membrane science have enabled removal of various substances from water. Many of the technical solutions have been developed in response to unexpected problems created by other technical advances.

Solutions have not been limited to water, however. Atmospheric pollution from automobiles has decreased dramatically in the last 25 years since the development and deployment of the catalytic converter (Box 2-1). Modern three-way catalysts can simultaneously reduce the concentrations of carbon monoxide, hydrocarbons, and nitrogen oxides in the automobile's exhaust stream. Also, new methods for remediation of contaminated soils have been provided by the selection of unique plants and microbes for this purpose.

---

[2]*Greatest Engineering Achievements of the 20th Century;* National Academy of Engineering: Washington, DC, 2000; *http://www.greatachievements.org/.*

BOX 2-1
Automotive Exhaust Emission Control

Automotive emissions have declined markedly over the past 15 years as illustrated in the chart below. With advances in combustion and after-treatment technologies, the period 1990-2005 will have seen an 80% reduction in hydrocarbon emissions and a more than 90% reduction in nitrogen oxide emissions. This remarkable improvement follows a comparable reduction in the preceding 15-year interval. The success of automotive emissions control has required a systems approach involving fuel quality, engine control theory, engine control computation power and software, engine control hardware, and catalytic converter technology.

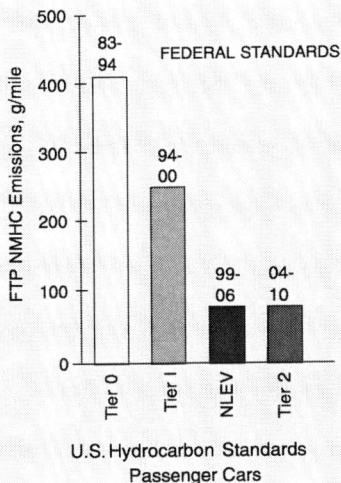

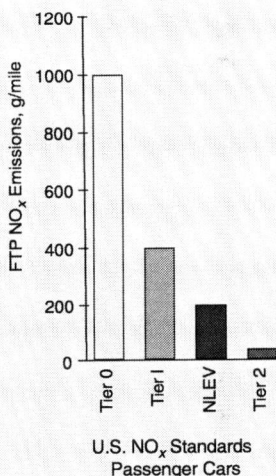

U.S. Hydrocarbon Standards
Passenger Cars

U.S. NO_x Standards
Passenger Cars

Catalytic Converter Technology for Today's Gasoline Engines

The three-way exhaust catalytic converter is used to complete the combustion of carbon monoxide and unburned fuel elements and to remove the NO and NO_2 produced during combustion. The primary components of the catalytic converter are the catalyst and its physical support. The catalyst is composed of a high-surface-area support that incorporates the primary catalytically active materials, typically mixtures of one or more of platinum, palladium, or rhodium (precious group metals, PGMs). The catalyst also contains promoters that improve the efficiency and stability of the PGM. Of these, the most important is cerium oxide, a

continued

BOX 2-1 Continued

promoter that is used to buffer small compositional swings in the exhaust composition within the catalytic converter.

Major improvements in hydrocarbon oxidation activity and overall thermal durability were achieved through the use of more palladium-rich formulations as lead disappeared from the fuel supply. Palladium is highly prone to lead-based deactivation. The dramatic increase in palladium in the late 1990s caused a significant supply-demand imbalance in the PGM markets in 2000, driving up palladium prices tenfold. The recent effort to reduce requirements for palladium without affecting hydrocarbon performance restored the market balance.

Additional improvement has come from 3-4 times faster warm-up of the catalytic converter to 250° C, the temperature at which it becomes active. The majority of vehicle emissions occur during this catalyst "light-off" period. Two features led to faster light-off. The catalyst was positioned closer to the engine to experience higher-temperature exhaust, and the thermal mass of the catalyst was reduced. This required catalyst materials with enhanced thermal durability and resistance to sintering and chemical poisoning from exhaust gas components. Neither thermal degradation nor chemical degradation is particularly well understood.

A key was the development of more thermally stable forms of cerium oxide. Unstabilized cerium oxide degrades rapidly at exhaust temperatures in excess of 800° C, and this degradation significantly reduces the efficiency of the catalyst for $NO-NO_2$ reduction. Greater thermal stability was achieved by forming solid solutions of stabilizers such as zirconium oxide and lanthanum oxide in cerium oxide. Hence, these new materials are the most significant enablers for the lower $NO-NO_2$ standards required for 2004 and beyond.

Reduced thermal mass came from changes to the catalyst support. The catalyst materials are deployed as a thin coating on a ceramic honeycomb substrate. The surface area for the deployment of the catalytic coating, and the mass of the substrate both contribute to the overall performance of the catalytic converter. The surface area was increased by reducing the thickness of the channel walls, thereby increasing the flow channel density and reducing the thermal mass of the substrate.

Catalytic Converter Technology: Lean-Burn Engines

The three-way catalyst works in a narrow range of air-fuel ratio, approximately the stoichiometric ratio. Hence, it is not effective with lean-burn engines that offer higher fuel efficiency at the expense of higher NO_x emissions. Catalytic systems under development for post-2004 U.S. NO_x emission regulations are based on NO_x traps, which adsorb NO_x as nitrates. These emission control systems are "active." They require peri-

continued

BOX 2-1 Continued

odic temperature increases and injection of reductant to regenerate the NO_x traps by release and catalytic conversion of the adsorbed nitrates to free the adsorption sites for further activity. Gasoline lean-burn engines can be controlled for periodic stoichiometric operation, but other strategies are required for lean-burn diesel engine systems. Prototype systems have used post-engine injections of urea or fuel as the triggering reductant with marginal success. Similarly, particles in diesel exhaust require filters where the particles are removed either continuously with a catalyst coating that promotes their oxidation or by periodic high-temperature excursions in engine conditions to incinerate particles trapped on a filter. Combustion in diesel engines is carefully controlled to balance NO_x and particulate matter (PM). Several prototype PM filter systems have been demonstrated successfully with diesel engines, but NO_x control remains a challenge for engine operation and catalyst technology.

The key chemical science challenge is kinetics. How does one obtain quantitative kinetic information for real, commercial, heterogeneous catalysts without resorting to time-consuming experiments? Similarly, simulations are required to design catalyst systems for performance at end of life, 150,000 miles. Simulations currently use empirically derived, aged-catalyst parameters. Driving vehicles to end of life to acquire data for variants of aftertreatment systems is laborious and does not provide sufficient statistical confidence. Hence, the science of accelerated aging and simulation tools for complex flow in chemically active heterogeneous systems is key to addressing challenges for bringing lean-burn, fuel-efficient engines to U.S. roads.

POLLUTION PREVENTION: SYNTHESIS, MANUFACTURING, AND PROCESS ADVANCES

Environmental contributions from the chemical sciences have not been limited to cleaning up existing problems. Major contributions have been also been made in *pollution prevention,* so that the undesired components are never generated in the first place. Some of the contributions include

• using life-cycle analysis to identify more clearly where and when environmental impacts occur in the cycle of raw material production and product manufacture, packaging, storage, use, and disposal;

• improving existing processes, such as the development of cleaner-burning fuels, for which less exhaust-stream treatment is needed;

- using substitutes—for example, replacing organic solvents with water or supercritical carbon dioxide to reduce emissions of organic solvents into the environment; other substitutes include replacement for heavy metals and chlorofluorocarbon (CFC) replacements as discussed in Box 2-2;
- enhancing energy efficiency and development of improved photovoltaic devices;
- developing degradable materials such as pesticides and polymers to reduce potential problems associated with persistent chemicals and materials;
- using biomass rather than petroleum as a feedstock for chemical processes; and
- employing atom economy as a strategy in chemical processes, thereby minimizing the amounts of waste for disposal.

Much of the contribution that has been made in these areas is encompassed by the phrase *green chemistry,* which is an approach to doing chemistry and chemical engineering that minimizes negative impacts on the environment through pollution prevention. It has been defined as the "design of chemical products and processes that reduce or eliminate the use and generation of hazardous substances."[3] Such processes continue to be developed, and they demonstrate that such technologies can be both economically and environmentally viable. Several speakers at the workshop presented examples of green process technologies (see Appendix D):

- Development of new catalysts eliminates carbon tetrachloride as a by-product in the production of phosgene from carbon monoxide and chlorine (U. Chowdhry);
- Manufacture of new high-solids enamel for automotive coatings reduces volatile organic compound emissions, reduces odor emissions by 86%, and reduces total raw materials use by 20% (U. Chowdhry);
- The use of a bio-based catalyst to make a key intermediate (5-cyanovaleramide) in the production of a new herbicide has increased the yield from 20% to 93% and greatly reduced the quantity of catalyst waste (U. Chowdhry);
- Polymerization of fluoroolefins in liquid and supercritical carbon dioxide eliminates the use of water and C-8 surfactant that has been identified as a persistent organic pollutant (R. Carbonell);
- The development of surfactants for CO_2 has made possible the commercialization of a dry cleaning process replacing perchloroethylene (R. Carbonell);
- The use of new catalyst in the process for manufacture of glyphosate (a herbicide with desirable environmental properties) results in the use of reduced waste, raw materials with lower toxicity and volatility, and lower energy consumption (M. Stern); and

---

[3]*http://www.epa.gov/opptintr/greenchemistry/*

• Genetic modification of soybeans, corn, canola, and cotton has produced glyphosate-resistant crops that lead to lower pesticide use, yield improvement, and improved water quality (M. Stern).

The Presidential Green Chemistry Challenge Awards Program[4] has recognized a wide range of accomplishments that include the following:

• Cost-effective production of 1,3-propanediol, a new feedstock for polyesters, using a genetically engineered fermentation pathway (DuPont, 2003);
• Polylactic acid, a polymer for fibers and packaging that is derived from renewable resources (Cargill Dow LLC, 2002);
• Waste-free manufacture of an environmentally friendly chelating agent (Bayer Corporation and Bayer AG, 2001);
• Enzymes and processes for large-scale organic synthesis (Chi-Huey Wong, 2000);
• A new catalyst for oxidation in pharmaceutical manufacturing that reduces chromium waste and solvent usage (Lilly Research Laboratories, 1999);
• The concept of atom economy (Barry M. Trost, 1998);
• Surfactants for supercritical CO_2 (Joseph M. DeSimone, 1997); and
• Manufacturing of polystyrene foam with CO_2 replacing CFCs as the blowing agent (Dow Chemical Company, 1996).

IDENTIFICATION OF PROBLEMS

Environmental problems cannot be solved if they have not first been detected. Many environmental problems have been chemical in nature, but they have occurred as unexpected consequences of other processes or developments. Discovering such problems—including identification of their underlying causes and elucidation of the details of the chemical processes involved—has relied heavily on the work of chemists and chemical engineers.

Atmosphere

Chemically related atmospheric problems have had high visibility in recent years. No matter how cautious one may be on the topic of global warming, it is clear that the greenhouse effect of carbon dioxide, methane, ozone, nitrous oxide, and the CFCs must be considered as a factor in global climate change. The molecular behavior of the greenhouse gases explain their ability to absorb infrared radiation from the earth and convert it to heat. Similarly, the photochemistry of chlorofluorocarbons (CFCs)—molecules originally believed to be completely benign—provided the explanation for stratospheric ozone depletion.

---

[4] *http://www.epa.gov/opptintr/greenchemistry/docs/award_recipients_1996_2002.pdf*

BOX 2-2 CFC Replacements [a,b]

For nearly half a century, chlorofluorocarbons (CFCs) were regarded as versatile wonder chemicals. Originally developed in the 1930s as non-toxic and nonflammable alternatives to hazardous chemicals such as ammonia and sulfur dioxide for refrigeration applications, in subsequent decades they found a host of applications ranging from cleaning and degreasing agents, to polymer foam blowing agents, to aerosol propellants. As the connection of these compounds to stratospheric ozone depletion was made, and the understanding of the chemical and physical processes behind this phenomenon developed, phase-out of the use and manufacture of these compounds was mandated, beginning with the Montreal Protocol of 1987 and its subsequent amendments.

In the decade and a half since, a number of application-specific alternatives have been developed. Cleaning and degreasing agents include liquid CO_2, citric acid, and other compounds for demanding applications such as microelectronics manufacturing. Blowing and propellant agents include CO_2 and butane, among others. Significant capture and recycling efforts have also been implemented, since release and large-scale accumulation of any chemical, CFCs or their replacements, in the environment is likely to have some negative impact, whether in the stratosphere or not.

The most visible consequence of CFC elimination for the consumer has been in refrigeration and air-conditioning applications. CFCs in late-model refrigerators, and home and automobile air conditioners have been supplanted by hydrofluorocarbons (HFCs) and hydrochlorofluorocarbons (HCFCs). CFC-12 (CF_2Cl_2) has been eliminated in automotive units in favor of HFC-134a (CF_3CH_2F). While this changeover has occurred in a remarkably short period, the magnitude of the scientific and technological challenges surmounted should not be understated. Ideal specifications for a CFC replacement include thermophysical properties of both the liquid and gaseous states that match those of the CFC in use, non-flammability, nontoxicity, compatibility with system components and lubricants, ease of manufacture, low cost, and of course, greatly reduced potential for depletion of stratospheric ozone. Meeting all of these challenges to come up with the alternatives that we use today required significant theoretical and experimental efforts. Advanced theoretical and computational tools were needed to calculate the thermodynamic properties of candidate CFC replacements and to model their transport, reaction, and ozone-depleting potential in the atmosphere. Experimental studies developed databases of HFC and HCFC property measurements, and created new catalysts and processes for manufacturing these compounds. New manufacturing facilities were constructed for commercial production.

Why is HFC-134a instead of CFC-12 in our cars and homes today?

The simple answer is that it has practically no impact (zero ozone depletion potential) on the stratosphere. The reason for this involves several different characteristics of the HFC molecule. First, it contains no chlorine, and therefore cannot release the principal ozone-depleting photochemical product of CFC-12, chlorine atoms, into the atmosphere. Second, because it incorporates hydrogen, HFC-134a is less stable than CFCs in the atmosphere. The strategy here may seem counterintuitive at first. If the deleterious effects of CFCs are due to their degradation products, one might attempt to make their replacements more stable and thereby reduce the release of such products. Yet the key is not just what and how much is released, but where. Part of the lower ozone depletion potential of HFC-134a is that it degrades sooner (i.e., at lower elevation) and does not reach the stratosphere to react with ozone there.

HFCs and HCFCs both act as greenhouse gases, so environmental concerns about those effects have not yet been resolved by using them as replacements for CFCs. However, their atmospheric lifetimes are only a few years, in contrast to those of the CFC, which may be many decades.

What lessons can we learn from this example? First, no compound, however wonderful, is likely to be entirely benign when released into the environment in large quantities and over a significant length of time. Sooner or later its effects will be felt because of its accumulation, appearance, and action in unexpected places (and often by unexpected mechanisms). The inertness of CFCs that made them appealing in a wide range of applications, including consumer products, allowed them to reach the stratosphere in significant quantities before degrading. Moreover the chlorine atoms released there act as catalysts for ozone destruction, greatly amplifying their impact. Thus, a second lesson has to do with the persistence and the action of compounds in the environment. It is desirable that these compounds exhibit limited lifetimes and that they and their degradation products not act by catalytic or self-replicating processes that amplify their environmental impact. Yet another lesson is that the process of reduction of the environmental impact of human creations is a continuous one. Although HFCs are in wide use today, some see these as merely the first step away from CFCs toward even more environmentally desirable alternatives. These and other challenges will continue to require the best efforts of chemists, chemical engineers, environmental scientists, atmospheric modelers, and investigators from a host of other disciplines, for (and on behalf of) generations to come.

---

[a]See *The Ozone Depletion Phenomenon,* Beyond Discovery, The National Academies, Washington, DC, 1996; *http://www.beyonddiscovery.org/.*
[b]Also see Box 3-2.

Chemists and chemical engineers have also contributed to the recognition of the importance of particulate matter in the atmosphere and the heterogeneous atmospheric chemistry that the particulate matter enables. Similarly, chemists and chemical engineers have provided the basic science that led to an understanding of photochemical smog formation and acid rain. This further helped to establish the role of biogenic emissions in the formation of smog.

Water

Many environmental problems in water result from chemical species that are present in only trace quantities. Consequently, chemical analysis and detection have made major contributions to discovering and understanding these problems. Examples include the problems of bioaccumulation of certain chemicals, persistent organic pollutants, pesticide residues, and the health effects of arsenic and lead as well as other trace metals.

Soil

Soil contamination also can be very difficult to detect. New analytical methods have made it possible to detect and analyze dioxins, polyaromatic hydrocarbons (PAHs), and polychlorinated biphenyls (PCBs). Analytical techniques have made it possible to detect the presence of these substances, determine whether remediation is needed, and evaluate the extent to which it has been carried out. Chemistry has provided or contributed to the remediation technologies that have been developed,

Chemists and chemical engineers have helped to develop a better understanding of interfacial processes. These govern the behavior of pollutants at the soil-water interface, and understanding them is essential to any remediation effort. Such interfacial processes can be of particular importance for the radioactively contaminated sites that were created over many years in the nation's nuclear weapons program.

INTERDISCIPLINARY DISCOVERIES

Chemists and chemical engineers do not work in isolation, and much of the work described here has involved collaboration with scientists and engineers in other disciplines. In many cases these collaborations have enabled particularly important advances in environmental science, as illustrated by the examples in the following list:

Biology

- Microbial in situ bioremediation and microbial community genomics
- Proteomics and metabolomics
- Polymerase chain reaction (PCR) and revolutions in molecular biology

Physics and Engineering

- Tools: new instrumentation; sensors, measurement systems and platforms

Materials Science

- New catalysts

Mathematics and Computer Science

- Multiscale computing (time and space)
- Bioinformatics (handling huge databases); analysis of high-throughput datasets
- Development and implementation of efficient molecular modeling software on advanced computers

Atmospheric Science, Meteorology, and Geology

- Biogeochemical cycles (carbon, nitrogen, etc.)
- Mechanisms and impact of contaminants

Many additional examples of collaborative work were brought up during the workshop, and these are summarized in Appendix G (Interfaces). It is frequently difficult to identify the particular scientific accomplishment that have led to environmental improvements or enhanced environmental understanding, in large part because environmental studies are inherently multidisciplinary. Consequently, cooperation across disciplines—as described above and in Appendix G—will continue to be necessary to fully understand and solve environmental problems, so the list also represents research opportunities for the future.

3

Challenges in Environmental Chemical Science

Chemical scientists and engineers have a particularly strong stake in the future of the environment. Many of the most important environmental threats are caused—or at least perceived to be caused—by release of undesirable chemicals into the air, water, or soil. In some cases, the source of such chemicals is natural, as in the highly publicized case of arsenic-contaminated groundwater in Bangladesh[1] and also in some parts of the United States.[2] Chronic exposure to small amounts of arsenic in drinking water increases a person's risk of cancer and other diseases.[3] High concentrations of arsenic found in the aquifers in Bangladesh and West Bengal pose serious threats to public health; estimates of population at risk run from 30,000,000 to a high of 80,000,000.[4] In other instances, human activity has been the origin of the chemical release. Some of the most important cases are also the most ironic—because the harmful effects on the environment were a direct consequence of a technological innovation that was intended to enhance environmental quality. Two compelling examples are provided by DDT (Box 3-1) and chlorofluorocarbons (CFCs, Box 3-2).

---

[1]*Arsenic Contamination of Groundwater in Bangladesh*; Kinniburgh, D. G.; Smedley, P. L., Eds.; Volume 1: Summary, BGS and DPHE, British Geological Survey Technical Report WC/00/19, British Geological Survey: Keyworth, UK, 2001.

[2]Focazio, M. J.; Welch, A. H.; Watkins, S. A.; Helsel, D. R.; Horn, M. A., *A Retrospective Analysis on the Occurrence of Arsenic in Ground-Water Resources of the United States and Limitations in Drinking-Water-Supply Characterizations:* U.S. Geological Survey Water-Resources Investigation Report 99-4279, 1999; *http://co.water.usgs.gov/trace/pubs/wrir-99-4279/.*

[3]*Arsenic in Drinking Water: 2001 Update,* National Research Council, National Academy Press, Washington, D.C., 2001.

[4]Smith, A. H.; Lingas, E. O.; Rahman, M. *Bulletin of the World Health Organization* **2000,** *78(9),* 1093-1103.

BOX 3-1
DDT and Malaria[a]

Malaria is a serious disease throughout much of the tropical world. In Ceylon (now Sri Lanka), for example, an epidemic in the mid-1930s resulted in nearly 50,000 deaths. The discovery by the Swiss chemist Paul Hermann Müller of the insecticidal properties of DDT in 1939 led to large-scale use of the compound around the world. The ability to control the mosquito population responsible for the spread of malaria led to widespread application of DDT in Sri Lanka starting in 1958. Nearly complete elimination of malaria resulted, with the rate of infection decreasing from more than a million cases each year to fewer than 20 cases reported in 1963. The battle appeared to have been won, and the spraying program was ended. This seemed to provide ample justification for the award of the 1948 Nobel Prize in Physiology or Medicine to Müller for his discovery.

However, the story does not have a simple and happy ending. Evidence was mounting that the mosquito population was becoming resistant to DDT. Elsewhere, concerns were developing that DDT might have harmful effects on other organisms. In the first example of an environmental contaminant that might have significant ecological effects, DDT was linked to defective eggshells in avian populations.

By 1967, a resurgence of malaria in Ceylon led to a renewed DDT program, but it was ineffective against a resistant mosquito population. In the late 1990s, the malaria infection rate increased to more than 200,000 cases annually.

The problems in this example have no easy solution. DDT use began as the apparent solution to a long-time scourge, but the success was short-lived. The challenge for chemical scientists now is to find new approaches to preventing or curing the disease—but without harmful effects elsewhere in the environment.

---

[a]See *http://plantpath.wisc.edu/ent371/Lect11%202002.pdf; http://info-pollution.com/ddtban.htm; http://users.rcn.com/jkimball.ma.ultranet/BiologyPages/I/Insecticides. html.*

How can environmental problems—such as those associated with arsenic, DDT, and CFCs—be solved? More specifically, how can the problems be resolved, how can release of pollutants into the environment be reduced or stopped, and how can harmful materials be removed from the environment? The complexity of the issue is even greater if the question is extended to, *Should* efforts be made to remove the material from the environment? Ultimately, this leads to the question, How can such problems be *prevented* in the future?

BOX 3-2
CFCs and the Ozone Hole[a,b]

Stratospheric ozone depletion is one of the best-established phenomena arising from anthropogenic influence on the global environment. As chlorofluorocarbons and other chlorinated and brominated substances are emitted into the atmosphere, those that are not subject to attack in the troposphere may reach the stratosphere where UV radiation breaks the molecules apart, releasing their halogen atoms. These halogen atoms initiate catalytic cycles that destroy stratospheric ozone; one chlorine atom can destroy as many as 100,000 ozone molecules before finally being removed from the stratosphere.

The introduction of these substances in the 1930s was viewed as an impressive step forward for human welfare. As working fluids for refrigeration, CFCs had ideal physical properties, and they were chemically inert and nontoxic. By replacing ammonia as the standard working fluid, a tremendous improvement in safety had been achieved. Once again, however, the story does not end on this happy note. A half-century later, it was found that these materials were the major contributors to ozone depletion in the stratosphere and the growing ozone hole over Antarctica, a phenomenon that threatened the well-being of life at the Earth's surface as a consequence of increased exposure to UV radiation.

While the inertness of CFCs was initially viewed as an asset, their long chemical lifetime enables them to diffuse into the upper atmosphere, where they undergo photochemical reactions when exposed to short-wavelength radiation from the sun. These reactions in turn lead to the destruction of stratospheric ozone and the weakening of the UV protection that ozone provides to Earth. For their work in elucidating these chemical pathways, the 1995 Nobel Prize in Chemistry was awarded to Paul Crutzen, Mario Molina, and Sherwood Rowland.

The ozone hole still exists, but the scientific discovery of the chemistry behind the problem has made a solution possible. CFCs and related persistent chlorinated and brominated organic materials are no longer being released in the massive quantities that were previously typical. Once again, the original problems remain—can we develop safe, nontoxic, and environmentally benign replacement materials for such uses as refrigeration, manufacturing, and fire suppression?

---

[a]See *The Ozone Depletion Phenomenon,* Beyond Discovery, The National Academies, Washington, DC, 1996; http://www.beyonddiscovery.org/
[b]Also see Box 2-2.

If science and engineering are to be employed in solving and preventing problems, then scientists and engineers will need to greatly increase the level of understanding of the relevant issues. Society will need to develop a far more comprehensive knowledge of the environment, from a global scale down to the chemical reactions that take place in air, water, and soil and within living organisms. New and more powerful methods for detecting and analyzing chemical substances will be needed, and it will be essential to develop a systems approach to the complex network of interacting chemical, physical, and biological processes that must be monitored and evaluated. Only with this greatly expanded knowledge will it become possible to fully protect, restore, and preserve our environment.

FUNDAMENTAL UNDERSTANDING

Human Influence on the Natural Environment

Evidence of anthropogenic influence on the natural environment is widespread: Examples include pesticide use (see Box 3-1), stratospheric ozone depletion (see Box 3-2), intercontinental transport of wind-borne dust and air pollutants, drinking water disinfection (see Box 3-3), and increasing levels of anthropogenic emissions into soils and groundwater.

Intercontinental transport of wind-borne dust has been observed for many years,[5] and satellites have tracked plumes of smoke from forest fires and biomass burning over thousands of kilometers.[6] More than two generations ago, it became recognized that lead, principally associated with the use of tetraethyl lead as a motor fuel additive, was being distributed globally. Likewise, long-range transport of chemical pesticides, such as DDT and other organic chlorine compounds, and the associated ecological damage have been a subject of study for decades. In the 1980s, ozone pollution was identified as not just an urban problem, and transport of ozone across national boundaries became an international issue in North America, Europe, and most recently, Asia. Surface measurements have shown that ozone pollution from North America is easily detectable 3000 km downwind from the North American source region.[7] Similar observations have been made of transport of Asian pollution across the Pacific.[8] Long-range transport of polychlorinated biphenyls (PCBs) and dioxins has been extensively documented. Recently, polybrominated diphenyl ether (PBDE) chemicals used as flame retardants in consumer products appear to be contaminating pristine areas of the Arctic

---

[5] Prospero, J. M.; Savoie, D. L. *Nature* **1989**, *339*, 687-689.

[6] Wotawa, G.; Trainer, M. *Science* **2000**, *288*, 324-328.

[7] Parrish, D. D.; Trainer, M.; Holloway, J. S.; Yee, J. E.; Warshawsky, M. S.; Fehsenfeld, F. C.; Forbes, G. L.; Moody, J. L. *J. Geophys. Res.* **1998**, *103*, 13357-13376.

[8] Jacob, D. J.; Logan, J. A.; Murti, P. P. *Geophysical Research Letters* **1999**, *26*, 2175-2178.

BOX 3-3
Drinking Water Disinfection and Disinfection By-Products[a]

Waterborne pathogens were responsible for tens of thousands of deaths during the first 150 years of U.S. history, with more than 50,000 deaths from typhoid fever as late as 1900-1904.[b] Worldwide, cholera outbreaks were eventually traced to contaminated water. Disinfection of water in the early 1900s with chlorine compounds and then with elemental chlorine nearly eliminated waterborne diseases such as cholera, typhoid, dysentery, and hepatitis in U.S. cities. Together with improvements in sewage treatment, this ranks among the most important contributions made by science and engineering to the improvement of human health.[c]

Disinfection with chlorine not only is effective against waterborne bacteria and viruses at the site of treatment but also provides residual protection by inhibiting microbial growth in the distribution system. While chlorine has been a true lifesaver, its use to treat drinking water also can generate disinfection by-products (DBPs) such as trihlomethanes (THMs) that may pose risks to human health. DCPs are of particular concern when treating surface water (from rivers, lakes, and streams) that is more likely to contain organic materials that can react with chlorine.

Under the requirements of the Safe Drinking Water Act and subsequent amendments, the Environmental Protection Agency (EPA) regulates DBPs in drinking water. A research effort has been established to facilitate decisions on drinking water safety by identifying and reducing amounts of DBPs found to be hazardous.[d] Currently, all public water

---

even more rapidly than either PCBs or dioxins.[9] The most comprehensive assessment to date of PBDEs in the breast milk of North American women indicates that the body burden in Americans and Canadians is the highest in the world, 40 times greater than the highest levels reported for women in Sweden.[10] Fluorinated organic compounds are globally distributed, environmentally persistent, and bioaccumulative.[11]

Water and Sediment Chemistry

Environmental chemistry has progressed significantly over the past four to five decades from a science that was concerned primarily with measurements of trends in the distribution of problematic species in the environment to the more

---

[9]Ikonomou, M. G.; Rayne, S.; Addison, R. F. *Environ. Sci. Technol.* **2002**, *36*, 1886-1892.

[10]Betts, K. S. *Environ. Sci. Technol.* **2002**, *36*, 50A-52A.

[11]Giesy, J. P.; Kannon, K. *Environ. Sci. Technol.* **2002**, *36*, 147A-152A.

systems that provide disinfection are required to test for THMs and meet the limits set by EPA. Water utilities have adjusted the type and amount of chlorine-based disinfectant used as well the site of application. In addition, the treatment process has been expanded to remove the naturally occurring organic matter that could react to produce THMs.

Chlorine is not the only means of disinfection available, but other methods can also produce toxic by-products. In addition, alternative disinfectants do not provide the residual protection offered by chlorine-based disinfectants, so they must be used in combination with chlorine. Drinking water treatment must satisfy the competing objectives of maximum microbial decontamination and minimum production of toxic by-products. This is a difficult task that will require research by chemists and chemical engineers in collaboration with a variety of other experts to continuously improve the safety and quality of the world's drinking water.

---

[a]Minnesota Department of Health fact Sheet: Drinking Water Disinfection By-Products, January 1998, *www.mrwa.com/mdh-drinkingwaterdisinfection.htm*.

[b]McTigue, Bill, "Hard to Swallow… The effects of water chlorination" *www.homeenv.com/Art_chlorination.htm*

[c]*Greatest Engineering Achievements of the 20th Century,* National Academy of Engineering, 2000, *http://www.greatachievements.org/greatachievements/ga_4_2.html.*

[d]Richardson, Susan "Disinfection By-Products of Health Concern in Drinking Water: Results of a Nationwide Occurrence Study, *www.epa.gov/ORD/scienceforum/PDFs/water/richardson_s.pdf.*

---

modern approach that seeks to understand processes on a fundamental chemical and physical basis. Accordingly, our ability to understand environmental processes hinges on our ability to recognize the factors responsible for the complex biogeochemical interactions that are collectively acting to modify the systems. Analytical chemistry plays a huge role in providing us access to methodologies that will enlighten our abilities to recognize the responsible factors. Therein lies the future in environmental chemistry, especially if one can apply developing methodologies to solve new problems. There are some critical thrust areas that have to be developed further. First is the need to understand chemical processes at a molecular level of detail. We need to move from empirical observations that feed well into large-scale models of transport, circulation, biodegradation, and other processes, to developing a fundamental understanding of the biogeochemical processes at a molecular level. It is only then that we can truly understand the factors governing such processes. We now have some very powerful analytical tools to do this, and we should strive to become proficient and knowledgeable in their application, especially with regard to solving some important environmental problems.

One problem that requires attention is developing a molecular-level understanding of how carbon turns over in soils, sediments, and waters. Much of our understanding of how our planet will cope with rising CO_2 levels and global warming trends is based on models that describe the behavior of carbon—its location, form (e.g., as carbon dioxide, carbonate, or organic plant matter), and rate of conversion through biogeochemical cycles—among various environmental compartments. One huge variable in predicting outcomes from models is the molecular manner in which carbon turnover occurs. The input data are based on uptake and evolution of CO_2 and nitrogenous components that provide only total rates. We really have little understanding of the factors that control turnover. Humic substances and their production from plant materials are important factors and we need to know more about the relevant biological and chemical processes down to the molecular level. Related to this is the nature of dissolved organic matter in natural waters—a huge reactive and storage reservoir for carbon. We also need to better understand the biological and physicochemical interactions that occur in response to global greenhouse gas augmentations. We also must not fail to consider effects that climate change might have on environmental chemical processes.

The major goals of environmental bioinorganic chemistry are to elucidate the structures, mechanisms, and interactions of important "natural" metalloenzymes and metal-binding compounds in the environment and to assess their effects on major biogeochemical cycles such as those of carbon and nitrogen. By providing an understanding of key chemical processes in the biogeochemical cycles of elements, such a molecular approach to the study of global processes should help unravel the interdependence of life and geochemistry on planet Earth and their coevolution through geological times.

[François Morel, Appendix D]

Dissolved organic matter (DOM) in water leads to the binding and transport of organic and inorganic contaminants, produces undesirable by-products through reaction with disinfection agents, and mediates photochemical processes. DOM is also a major reactant in and product of biogeochemical processes and controls levels of dissolved oxygen, nitrogen, phosphorus, sulfur, numerous trace metals, and acidity.[12] DOM can range in molecular weight from a few hundred to 100,000 Da, which is in the colloidal size range, and generally has similar characteristics to humic substances in soil. Moreover, contaminants of a bewildering array are

---

[12]Leenheer, J. A.; Croué, J.-P. *Environ. Sci. Technol.* **2002**, *36*, 19A-26A.

being found in aquatic environments. These include detergents, disinfectants, insect repellents, fire retardants, plasticizers, reproductive hormone mimics, steroids, antibiotics, and numerous other prescription and nonprescription drugs.[13]

Another important area is understanding the fate and transport of anthropogenic chemicals in soils and sediments. Organic chemicals, including pharmaceuticals, fertilizers, herbicides, and pesticides, are at the top of the list. It is important to develop ties among environmental chemists and engineers who model processes associated with fate and transport. Most models employ empirically derived parameters for the kinds of interactions that accelerate or retard such processes. These models are often poorly developed for lack of a better understanding of molecular-level processes. We need to know how contaminants are hydrologically transported in a medium where they continually interact at the molecular level with DOM and with mineral, biological, and organic phases of soils and sediments.

The occurrence and mobility of harmful chemical substances, whether of natural origin or anthropogenic contaminants, in the subsurface environment pose both an intellectual and fundamental scientific challenge and practical concerns for the use and management of groundwater resources. The chemical sciences offer powerful approaches toward understanding and mitigating the problems of groundwater contamination. Society has benefited and will continue to benefit from this important application of chemistry to environmental problems.

[Janet Hering, Appendix D]

Another area that deserves attention is *natural attenuation*—the diminution of pollutants in soil and groundwater by such natural processes as adsorption, dilution, dispersion, chemical or biological degradation, radioactive decay, and vaporization that take place without human intervention. This work has gained popularity in recent years, mainly because it is being used to justify corporate and governmental decisions. We need to evaluate the process in depth, again at the molecular level. The Earth system may well be capable of remediating itself, but we need to know the time scale and the outcome associated with such an approach. Chemistry plays a monumental role along with biology and environmental engineering in developing the crucial experiments to evaluate the fate of individual contaminants. Once the underlying science is well-understood, it may be possible to develop low-cost enhancements or acceleration of the natural processes.

---

[13]Erickson, B. E., *Environ. Sci. Technol.* **2002**, *36*, 141A-145A.

Chemistry has and should continue to play a leading role in developing an understanding of the toxicological effects of our industrialization—effects resulting from petrochemicals, specialty chemicals, nuclear wastes, natural geochemical hazards, and, foremost, from the processing of our drinking water and foods. Developments in analytical methodologies and approaches can significantly extend our knowledge of the harmful effects of anthropogenic chemicals in the environment. Of specific concern and importance is the molecular-level relationship between bioavailability and the chemical speciation of various chemicals of environmental concern. Currently, there is great interest in understanding how organisms utilize, in either a beneficial or a deleterious fashion, specific compounds in their surroundings. The specific form of the compound is crucial to understanding whether it is harmful, beneficial, or benign. The flurry of activity in bio-inorganic chemistry is central to this issue, but the importance also extends to the speciation of organic compounds, including organometallic compounds. Not only does chemistry play an important role in characterizing the compounds of interest (a challenge for analytical chemists who often must do this at subpicomolar concentrations), but it is central to understanding the processes by which bioavailable forms become incorporated into cellular structures in organisms.

Gas-to-Particle Conversion and Combustion Aerosol Formation

The major processes for creating atmospheric fine particles (diameter < 2.5 μm) are combustion and gas-to-particle conversion (GPC). Whereas combustion particles are emitted directly to the atmosphere (primary aerosol), gas-to-particle conversion refers to the chemistry that leads to particulate matter by converting volatile gases into condensable substances under atmospheric conditions. Gas-to-particle conversion leads to an increase in the mass of preexisting particles and under some circumstances may lead to the creation of new particles. Particulate material produced by GPC is referred to as secondary aerosol.

Understanding GPC entails identifying precursor gases, elucidating the chemistry that converts them (in the gas phase, on surfaces, or in solution) into condensable species, and determining the processes by which those species are then converted into particulate matter (e.g., by nucleation, condensation, or direct production on an existing particle). One can make a distinction between GPC processes that lead to new particles where gas-phase chemistry produces supersaturated conditions followed by nucleation into a single-component or, more likely, a multicomponent condensed phase and processes that chemically age a preexisting aerosol by heterogeneous or multiphase reactions and condensation.

An understanding of GPC (both nucleation and growth) requires knowledge of which of the possible species participate in nucleation, their concentrations, and their thermodynamic and nucleation properties. For the atmosphere, one needs the concentrations not only of precursor gases but also of the oxidants, ozone, hydroxyl radicals, and nitrate radicals that initiate the process. Important gas precursors that have been identified include sulfur species such as sulfur dioxide and dimethyl sulfide, volatile organic compounds (VOCs) such as aromatics in urban areas and monoterpenes in forested regions, and ammonia and nitric acid. Reactions of these species lead to low-volatility products such as sulfuric acid, ammonium sulfate, ammonium nitrate, and multifunctional organic compounds containing acid, carbonyl, hydroxyl, and nitrate groups.

Formation of combustion particles also involves nucleation and condensation of vapors, although the processes occur at elevated temperatures inside the combustion source and during cooling of the plume. Like secondary aerosols, combustion particles have a major semivolatile component composed of sulfates from sulfur dioxide oxidation and organic oxidation products, and of unburned fuel and oil as well. Furthermore, they contain a large non-volatile component consisting of soot, metals, and metal oxides.

The most important problems involving fine particles involve their potential impacts on global climate and human health. Climate effects can occur through direct scattering and absorption of radiation and by altering cloud radiative properties and lifetimes through the action of particles as cloud condensation nuclei. The mechanisms by which particles impact human health, such as respiratory and cardiovascular function, are not yet fully understood, particularly in relation to particle size. Nevertheless, epidemiologic studies have been sufficiently convincing to result in implementation by the EPA of new air quality standards for fine particulate matter.

The properties of particles that determine their impacts on both global climate and human health include number, concentration, size, composition, mass, and surface area. An aerosol may be chemically inhomogeneous from particle to particle, resulting from a mix of processes. Consequently, characterization of individual atmospheric particles, rather than of bulk particulate matter, is most desirable for sorting out these details. In addition, there is a need to characterize the chemical structure within particles. Whether individual particles are chemically homogeneous or have organic coatings or inorganic incrustations, for example, will affect their radiative properties, atmospheric removal, heterogeneous reactions, and role as nuclei for cloud droplet formation and growth, as well as their health consequences.

Although environmental chemists have traditionally flourished in the realm of atmospheric processes, some new discoveries have confounded explanation. For example, scientists have begun to recognize the important role of black carbon (small particulate matter formed as a by-product in combustion of fossil fuels) in greenhouse trend reversal. The absorption of sunlight by aerosols containing black carbon can result in simultaneous warming of the atmosphere and cooling of the Earth's surface. However, evaluating the role of black carbon is difficult due to the lack of good measurement methodologies. The evolution of arctic smog during the summer season is an enigma not easily understood, but it is clear that understanding the molecular-level chemistry and photochemistry in aerosols is crucial.

The utilization of mass-independent isotopic measurements of atmospheric, hydrospheric, and geologic species has advanced understanding of a wide range of environmental processes. The future development of the utilization and understanding of this new technique clearly will have numerous applications that should, and will, be advanced. Issues in climate change, health, agriculture, biodiversity, and water quality all may be addressed. Simultaneous with the acquisition of new environmental insight will be the enhancement of the understanding of fundamental chemical physics.

[Mark Thiemens, Appendix D]

Finally, there is an important role that environmental chemists can play in homeland security,[14] especially in areas such as radioactive contamination; deliberate contamination of water, food, air, and soil; and detection of potentially harmful devices. It is likely that the future of our society depends on our ability to respond to the effects induced by acts of terror. Environmental chemists, employing state-of-the-art analytical tools, can play a leading role in prevention, mitigation, and prediction of harmful effects. Dual-use technologies will have applications in monitoring environmental change, alerting society to homeland security threats, and characterizing dangerous agents. The integration of environmental measurements in a network could also identify when and where something unusual occurs.

Putting It All Together: Understanding Biogeochemical Cycles

The real grand challenge is to understand fully the operation of biogeochemical cycles and the implications of human use of chemical feedstocks. Most of today's environmental issues evolved from ignorance or disregard of the fates of the chemical by-products of human endeavors, and a full understanding of the complex interrelationships will be a necessary foundation for resolving the issues. This will be a difficult task, because the biogeochemical cycles cross a variety of boundaries. They involve the different scientific disciplines of chemistry, biology, and geology; they cover water, soil, and air; and they include different scales from nanometers to kilometers.

---

[14]*Challenges for the Chemical Sciences in the 21st Century: National Security & Homeland Defense,* National Research Council, The National Academies Press, Washington, D.C., 2002.

INSTRUMENTATION AND RESEARCH TOOLS

Chemists and chemical engineers are uniquely situated to help advance understanding of the environment, largely because of their molecular approach. Moreover, tremendous progress has been made in the levels of precision and accuracy that chemical measurements can provide. This has been made possible by the development of fundamental theory and a good understanding of the thermodynamics, kinetics, and transport properties that govern the interactions of molecules. From a chemical perspective, the fundamental challenges in environmental problems arise from expanding this understanding to enormously greater scales of time and space while maintaining an accurate representation of both physical and chemical processes. A few years ago, even contemplation of such an achievement was hopeless, but recently the combined advances in analytical instrumentation and computational tools have allowed some of the challenges to be addressed. The value of instrumentation for measurement, coupled with robust computational and mathematical tools for modeling, can hardly be overestimated. Without measurement and modeling, understanding of environmental phenomena cannot be achieved. Without modeling, experiments cannot be designed and tested.

Chemists and chemical engineers use three basic tools to measure and describe the chemistry that is taking place in our environment:

1. laboratory measurements and devices to simulate and characterize that chemistry;
2. theoretical and computational tools to check experimental results, extrapolate, and codify what we know in environmental models; and
3. field observations and experiments to make direct environmental measurements and learn the impact of human activities.

The components of our environment—atmosphere, water, and soil, as well as the biological systems with which they interact—are all critically important to our well-being and are markedly undersampled. Much remains to be learned about the behavior and effects of aerosols, micrometer-size particles, and trace molecules in air and water. Understanding such phenomena is important to human health, ecology, and climate. Moreover, understanding the chemical basis for their biological effects is within our grasp. We are developing the capabilities for measuring the environment on a worldwide basis, and the global migration of pollutants is being documented for the first time. Success in these endeavors will depend on rapid advancement in measurement and modeling.

In measurement science, the desired signal is often very small and may be embedded in highly variable samples. Environmental samples are seldom controlled. They are likely to contain complex mixtures of compounds in a wide variety of matrices that can further complicate the sampling problem. Furthermore, knowing only the composition is inadequate. To fully comprehend the problem, one must have information on horizontal and vertical fluxes of the species being studied. The sources, sinks, and fluxes between reservoirs must also be understood, including their variation over time in response to atmospheric and terrestrial perturbations.

As noted by Dellinger (Appendix D), combustion sources are responsible for a major fraction of air pollution problems. This understanding began with the recognition that NO_x and organic materials from combustion sources are the primary source of photochemically induced air pollution, and it was also found that SO_x and NO_x emissions are responsible for acid rain. Particulate emissions, initially considered to be mainly a nuisance, are known to be primarily responsible for atmospheric hazes. Through use of refined analytical measurements, biological assays, and epidemiological studies, it was gradually recognized that particulate matter had human health impacts, first as a lung irritant, then as a source of PAHs, and now a source of other, yet undefined, biologically active pollutants contained primarily in the carbonaceous fraction. It was also found that PAHs are carcinogens, and combustion is again their principal source in the environment.

Although combustion and thermal processes are necessary to provide for the essential needs of our existence, they are intrinsically "dirty" and emit a variety of air pollutants. Some of these pollutants are well known, well understood, and subject to significant control.

However, combustion is a complex process that results in formation of many pollutants that are not well characterized as to their nature or origin. As a responsible society, it is incumbent upon us to examine these issues, determine their importance, and endeavor to eventually resolve and address each of them.

[Barry Dellinger, Appendix D]

It is now generally accepted that combustion is the almost exclusive source of the carcinogen and endocrine-disrupting chemical family of polychlorinated dibenzo-*p*-dioxins and polychlorinated dibenzofurans (PCDD/F or *dioxins* for short). Dioxins are part of a broader class of chlorinated organic air pollutants that are produced by most combustion sources. All that is necessary are a small source of chlorine (that may even be in the combustion makeup air), a catalytic metal, and a hydrocarbon.

Extensive research, enabled by improved analytical measurements, has been carried out on the mechanism of formation of dioxins and other chlorinated organics. Researchers discovered that many combustion-generated pollutants actually are formed after the flame zone not only by surface catalyzed reactions but also in the gas phase by high- and medium-temperature thermal processes. These discoveries were highly dependent on improvements in measurement tools and better mathematical models.

Measurement Tools

In carefully controlled laboratory environments, it is possible to use such methods as mass spectrometry and laser spectroscopy to make sophisticated measurements to reach the ultimate limit of detection—the detection and characterization of single molecules. However, these limits are not easily achieved with environmental samples, where sampling and separations challenges are as important as detection sensitivity. Moreover, laboratory investigations in controlled environments by skilled investigators are important, but they cannot approach the needs for environmental sampling in real time.

Atmospheric measurements are also challenging because they must deal with low to extremely low concentrations of trace chemical species. The major components (>99.999%) of the lowest portions of the atmosphere (the troposphere up to ~10 km in altitude and the stratosphere between ~10 and ~50 km) are molecular nitrogen, molecular oxygen, argon, water vapor, and carbon dioxide. Chemists will recognize that all of these species are very stable, strongly bonded molecules or atoms that are essentially inert gases at normal atmospheric temperatures (190-310 K). Indeed, without solar photons to break up selected molecules, atmospheric chemistry would be very dull indeed.

Atmospheric chemistry is dominated by trace species, ranging in mixing ratios (mole fractions) from a few parts per million, for methane in the troposphere and ozone in the stratosphere, to hundredths of parts per trillion, or less, for highly reactive species such as the hydroxyl radical. It is also surprising that atmospheric condensed-phase material plays very important roles in atmospheric chemistry, since there is relatively so little of it. Atmospheric condensed-phase volume to gas-phase volume ratios range from about 3×10^{-7} for tropospheric clouds to ~3×10^{-14} for background stratospheric sulfate aerosol.

[Charles Kolb, Appendix D]

The challenges in monitoring the environment are enormous. The surface area of the Earth is about 500 million square kilometers, two-thirds of which is ocean. The relevant volume of the atmosphere—the troposphere where we live and the stratosphere above that—is 25 billion cubic kilometers, and its composition is constantly changing. Many millions of dollars are spent for instruments on planes, satellites, and ground stations, just to obtain meteorological data for weather forecasting. To study chemicals in the atmosphere—as gases, mists, and particles, with their fluxes and phase changes—presents a monumental challenge for environmental and analytical chemists as well as those who will develop the necessary instrumentation.

How can we meet this challenge with the tools we have or the tools we might be able to develop? Three strategies are available:

1. *Very fast sensors.* By mounting such sensors on mobile platforms and moving them to sample over a range of space and time, it would be possible to understand how a system is actually responding.

2. *Remote sensing.* This works well for the atmosphere, sometimes for the oceans, and for the first 10 cm of the soil. The National Oceanic and Atmospheric Administration (NOAA) and National Aeronautics and Space Administration (NASA) have enormously successful, albeit expensive, satellite programs for monitoring and understanding the Earth's environment on an ongoing basis.

3. *Sensor arrays.* By linking a large number of sensors in an intelligent way, the value of the information that is gathered can be greatly amplified. However, implementation of such an approach will require that both the sensors and the necessary communications system be inexpensive.

One possible solution to real-time global sampling would be a combination of widely distributed robust measurement systems; highly mobile sensing platforms; and satellite, shipboard, and aircraft-mounted instruments. These remote sampling stations and platforms would have to be backed up by high-performance laboratory-scale instrumentation to verify field results. In addition, they would have to be connected via information technology networks to central processing computers, and these large data arrays would require extensive processing using chemometric approaches.

The information needed from the chemistry community on combustion particles and GPC in order to significantly advance understanding of the impact of aerosols on global climate and human health, and possibly ameliorate these effects, includes the following:

• *Thermochemical data from either measurements or computations on organic compounds, including vapor pressures and solubility in organic and aqueous salt solutions.* Data on cluster properties of organics and the sulfuric acid-water-ammonia system are necesary for understanding nucleation.

• *Heterogeneous reactions.* Knowledge of atmospherically relevant heterogeneous reactions is far from complete. Important reactions probably still remain to be identified and their rates and mechanisms determined. Just as ignorance of heterogeneous chemistry contributed to the failure of stratospheric ozone models to anticipate the formation of the antarctic ozone hole, much still is to be discovered and learned about the role of heterogeneous reactions in the troposphere.

• *Advanced methods for particle analysis.* The state of the art in quantitative methods includes semicontinuous analysis of major inorganic cations (NH_4^+, Na^+, K^+, etc.), anions (SO_4^{2-}, NO_3^-, Cl^-, etc.), and organic acids using particle collection and ion chromatography, and real-time particle mass spectrometry using thermal desorption-electron ionization. These methods do not analyze single particles or refractory materials. A number of instrument designs based on laser-vaporization–time-of-flight mass spectrometry are now available for analyzing the size and composition of single aerosol particles (including refractory components) in real time. Although these instruments are quite powerful, their limitations include the following: (1) they are generally limited to analysis of particles larger than ~50 nm in diameter, (2) they measure the composition of an uncontrolled and unknown fraction of a particle, and (3) measurements are not quantitative.

For public drinking water, groundwater is a significant resource. In the United States, 46% of drinking water comes from groundwater resources, and 54% comes from surface water. Groundwater is considerably less vulnerable to pathogens, but its greater opportunity to interact with soil minerals allows various dissolved species to accumulate in the water supply. The problems associated with lead and arsenic are well known, but to fully understand the evolution of groundwater composition, it will be necessary to obtain a much greater insight into the biogeochemical processes by which various chemical constituents are partitioned between mobile and immobile phases in the aquifer.

Analytical needs and opportunities in this area are challenging, particularly for pollutants of emerging concern such as endocrine disrupting compounds and pharmaceutical derivatives. Direct spectroscopic measurements of the sort than can be used for atmospheric measurements are not usually applicable. Sample collection, preparation, and analysis typically have been carried out in separate steps. Consequently, the development of techniques for in situ measurement capability, remote sensing and detection, and sensors for monitoring in soil and water would afford significant progress.

Analytical Instrumentation

A broad range of developments, generally described as *laboratory-on-a-chip* technologies, have been described and are being explored in several laboratories.[15] A variety of lab-on-a-chip combinations can be used to multiplex a variety of separations with multiple detectors. Techniques such as molecular imprinting offer considerable promise for reducing the cost of these devices while maintain-

[15]Ramsey, J. M. *Nature* **1999**, *17*, 1061-1062.

ing high sensitivity in a robust and compact package. More sophisticated versions of the devices have used micro-versions of laser-induced fluorescence and microscale ion-trap mass spectrometers. The latter two detection methods are among the most versatile and sensitive means for characterizing molecules. Laboratory versions are capable of both single-molecule sensitivity and a very high degree of specificity. Miniaturized versions of such devices lose some but not all of their versatility and sensitivity. Soon these approaches will emerge as robust, field-deployable measurement tools.[16]

A particle-sampling mass spectrometer could provide a useful approach to measuring inhalation exposure to pollutants in a wide variety of environments. One example[17] employs an aerodynamic lens that samples very fine particles and creates a beam that can be modulated to give a crude time-of-flight mass distribution. The particles impinge on a hot surface, causing vaporization of constituents that can be ionized (by electron impact or photoionization) and subjected to mass analysis by any of several kinds of mass analyzers.

Combinations of spectroscopies, such as nuclear magnetic resonance (NMR) and fluorescence or mass spectrometry to obtain visual and time evolution information (NMR) and high specificity and sensitivity (optical and mass spectrometry) are also highly promising new approaches for measurement science. However, it is beyond the scope of this report to list the many other recent advances in measurement sciences that are highly promising and clearly applicable to environmental problems. Many are at an early stage of development, and progress will require cross-disciplinary teams of chemists, physicists, engineers, computer scientists, and instrument specialists. Since the potential market is undefined and the number of people with the requisite skills is limited, ways and means must be devised to bring the best ideas of academia, industry, and federal laboratories to bear on this problem.

The cavity ring-down laser (see J. Anderson, Appendix D) is an exciting new development with great potential. One method[18] uses highly reflective mirrors mounted about a meter apart; the laser beam transits the cavity more than a million times, giving an effective detector path of about 10 km. Such complex and sensitive devices are not yet ready for widespread implementation, but their use in aircraft has been demonstrated. Reducing their size, incorporating commercial infrared lasers, and downsizing them into a shoebox-size package will be the next challenge.

---

[16]Patterson, G. E.; Guymon, A. J.; Riter, L. S.; Everly, M.; Griep-Raming, J.; Laughlin, B. C.; Ouyang, Z.; Cooks, R. G. *Analytical Chemistry* **2002**, *74*, 6145-6153.

[17]Jayne, J. T.; Leard, D. L.; Zhang, X.; Davidovits, P.; Smith, K. A.; Kolb, C. E.; Worsnop, D. R. *Aerosol Science and Technology* **2000**, *33*, 49-70.

[18]O'Keefe, A.; Scherer, J. J.; Paul, J. B.; Saykally, R. J. *Cavity-ringdown Spectroscopy: an Ultratrace-Absorption Measurement Technique;* Busch, K. W.; Busch, M. A. editors; ACS Symposium Series 720; American Chemical Society, Washington, D.C., 1999, pp. 71-92; Provencal, R. A.; Paul, J. B.; Chapo, C. N.; Saykally, R. J. *Spectroscopy* **1999**, *14*, 24.

Individual instruments will not be used to solve most environmental problems. It will be necessary to develop suites of instruments that can be cross-calibrated and cross-correlated to show that the same part of the environment is being examined and that the measurements are taking place at the same time. It is likely that in situ and remote sensing will be used concurrently, while the resulting data are evaluated in a model in real time. For example, impressive and expensive mobile measurement systems were used in evaluating the problem of stratospheric ozone depletion by simultaneously measuring several different radical species (halogen, hydrogen-oxygen, and nitrogen oxide) in real time, with 1- to 5-s resolution.[19] Similarly, in Mexico City a mobile laboratory was used for real-time measurements of fine particle and gas-phase species, looking at emission sources, process studies, mapping, and understanding how one part of the city differs from another both in emissions and the chemical reactions taking place in the atmosphere.[20]

Additional analytical capabilities needed include the following:

• Instruments for remote sensing that are robust, portable, and miniaturized.
• High-throughput instrumentation for sampling and analyzing large numbers of samples.
• Instruments and methods that can measure single-particle composition in real time down to sizes of fresh nuclei (~1 nm).
• Instruments and methods that can in near real time characterize more fully the speciated organic composition of secondary and combustion aerosols and that of the gas phase. In conjunction with laboratory studies, one may hope to use these techniques to elucidate the pathways and connect precursor volatile organic compounds to the nature of particulate matter.
• Instruments and methods that can provide information on the composition and structure of particle surfaces. These would have to be real-time or near-real-time measurements because surfaces of collected particles would be prone to alteration.
• Improved methods for collection of semivolatile compounds. State-of-the-art denuder systems are still prone to adsorption-desorption artifacts. Because of the high costs of particle mass spectrometers, most studies of organics will probably rely on particle collection with off-line analyses by gas chromatography-mass spectrometry (GC-MS), so reliable collection methods are important.
• Instruments and methods for measuring atmospheric particle water content, which is quite difficult because of the volatile nature of water.

[19]Brune, W. H.; Anderson, J. G.; Chan, K. R *J. Geophys. Res.* **1989,** *94(D14),* 16,649-16,663.
[20]Jayne, J. T.; Leard, D. C.; Zhang, X.; Davidovits, P.; Smith, K. A.; Kolb, C. E.; Worsnop, D. R. *Aerosol Sci. Technol.* **2000,** *33,* 49-70.

   • Instruments and methods for measuring particle pH. In light of recent studies indicating that acid-catalyzed particle-phase polymerization reactions may be important in secondary organic aerosol formation, ambient particle pH must be known to evaluate and model such reactions.

   • Identification of chemical tracers for the wide variety of organic combustion and secondary aerosol sources, which can be used to identify and quantify source contributions to ambient aerosol. For the health effects community, it would especially valuable to have simple indicators of sources, which need not be highly accurate but can be easily measured and used to correlate source contributions with health criteria.

Computing Tools and Applications

   The introduction last year of the Earth Simulator in Japan, which leapfrogged development of high-performance computing in this country, has provoked considerable discussion among academic and federal agencies. This computer is a general purpose, vector computer that is applicable to a wide range of problems in science and engineering, and it is more than a hundred times faster than the fastest computer accessible to nondefense scientists in the United States. David Dixon (Appendix D) described a high-performance computer with almost one-third of the computing power of the Earth Simulator that should be readily adaptable to general environmental applications.

   With such higher-performance computers, accurate quantum mechanical calculations should be possible with heavy elements for which relativistic corrections are significant, accurate zero-point energies can be calculated to improve the accuracy of thermodynamic calculations dramatically, and rate equations can be used to search more accurately for transition states and to study interfacial reactions (including those at the cellular interfaces of biological systems). The importance of improving accuracy in these basic calculations can hardly be over-estimated. As an example, consider an error of 0.2 kcal/mol in the activation energy to form the water dimer. When propagated during a calculation of nucleation phenomena the initial error could be expanded by 40 orders of magnitude, leading to an error of 10^{12} in rate constants for nucleation.

   Progress using this class of computer requires teams of experts in computational science, computational chemistry, applied mathematics, and software engineering to adapt and effectively use these tools. A major role for chemists is defining the problems to be tackled with such tools. Modeling tools—many of which require development—will be essential to traverse all relevant length and time scales, and computation will have to be coupled with measurement to confirm models and make predictions. With success in this realm, modeling increasingly will replace experiments that are too difficult, too dangerous, or too expensive.

   Data storage and integration are major challenges for computer science and mathematics specialists. These challenges must be addressed before the sophisticated instrumentation described above can be employed with optimal effectiveness. Thus, except for demonstration units, few such instruments exist. Connection to sophisticated computer networks and informatics processing are critical requirements for converting raw data into useful information.

Modeling, Simulation, and Computational Chemistry[21]

What can theory and simulation accomplish? They greatly enhance our understanding of known systems, providing qualitative and quantitative insights that can enable new experiments or new systems. Computational chemistry has the goal of providing quantitative results that can eliminate the need for experiments that are too difficult, dangerous, or expensive or can extend into temporal or spatial domains where the necessary experimental tools are not available. Calculations have become increasingly reliable, providing valuable methods for solving Newton's laws of motion for molecular dynamics and Schroedinger's equation for electronic motion. Extending these capabilities and solving the nuclear motion problem will be an important challenge in the coming decades. Progress will enable calculation of molecular structures and energetics, reaction equilibria, substitution effects, spectroscopic properties, rates of reactions, and reaction mechanisms.

Computational chemistry can play a key role in advancing the scientific enterprise. It can provide the data input for many larger, more complex models and provide us with unique insights into molecular behavior so that we can design and construct new molecules for specific tasks. Computational chemistry has become an established tool in the chemist's toolbox and is being used in broad areas of chemistry to replace experimental measurements and to provide us with improved understanding of molecular behavior. Computation will be the major tool that will enable us to cross the many temporal and spatial scales that characterize environmental science.

[David Dixon, Appendix D]

When applying computational chemistry to complex environmental problems, major challenges are encountered in scale—both in time and in space. This is illustrated in for atmospheric chemistry in Figure 3-1. Current methodologies

---

[21]Many of the relevant issues are covered in one of the other reports in this series, *Challenges for the Chemical Sciences in the 21st Century: Information and Communications,* National Research Council, The National Academies Press, Washington, D.C., 2003.

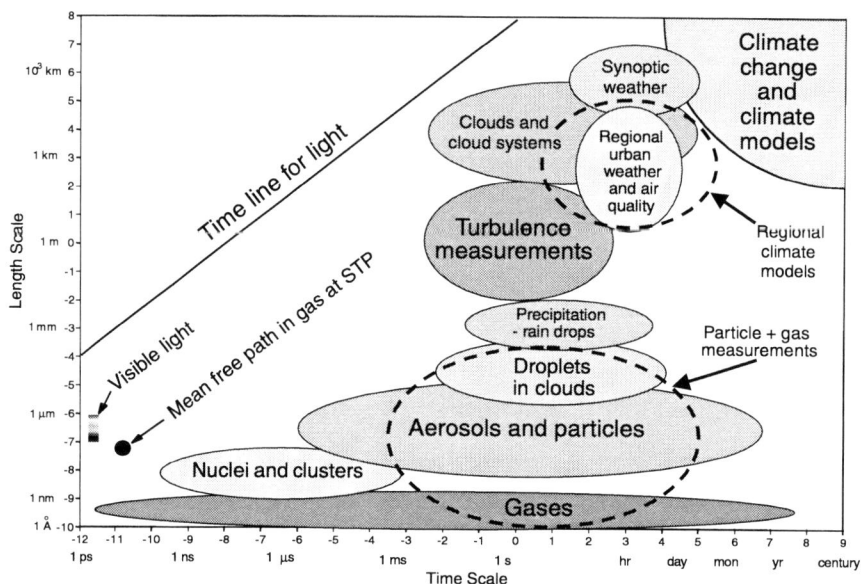

FIGURE 3-1 Dimensions of integration and the problem of scale as illustrated for atmospheric chemical calculation.

do not connect the molecular scales with all of the subsequent scales above them. We lack the tools, theories, and methods to make the connections. Increasing the scale of the models will require increased computing power, and advances will rely on collaboration with mathematicians and computer scientists. These problems of scale are analogous to those encountered in a chemical plant, where modeling and simulation must span the range from molecular processes and chemical kinetics through process optimization. Applications that range from climate modeling to bioinformatics and genomics also will require new developments in the analysis and manipulation of increasingly massive datasets.

Many important environmental questions or problems will necessitate computational study if they are to be solved. Such problems include

- the urban-to-regional migration of nitrate, sulfate, heavy metals, and organic soot into the broader environment—and the associated public health issues;
- forecasting climate change—and testing the forecast in a way that is acceptable to decision makers in science and in public policy;
- ultraviolet dosage and its relationship to ozone depletion;
- structure-toxicity relationships;
- forecasting adaptation to global climate change;
- chemical behavior and trends in oceans and estuaries;
- molecular origins of toxicity (including gene-toxicant interactions);

- integration of massive datasets dealing with many variables;
- systems modeling of the environment, accounting for coupled transport processes and reactions in air, water, soil, and their interfaces; and
- extrapolation of laboratory results over orders of magnitude in length and time.

Each of these problems, many of which are linked, must be attacked with both observations and modeling. Consider for example the first three entries in the preceding list. The issue of global warming depends on carbon sources and sinks, and it ties back directly to the questions of nitrate, sulfate, heavy-metal, and organic-soot emissions—from regional to global extent. The experimental requirements for addressing the three problems require high spatial resolution of fluxes, isotopes, and reactive intermediates. The details of molecular fluxes—how they interact with the boundary layer and how the boundary links into the free troposphere—have not yet been addressed properly.

An entirely new level of sophistication—not only in experiments but also in modeling—will be required for particles, aerosols, and the associated radiation field sets. New mid-IR laser-based instrumentation and use of long-duration balloons have helped make major advances in observations. The balloons can sit in the upper stratosphere and then be lowered to the lower stratosphere with power from fuel cells and solar panels. The modeling elements are equally important: it is necessary to test the model and its validity, and the model must link the measurements. The observations must be linked to trajectories, the trajectories must be initialized, and sources of specific chemicals must be identified along with the positions of those sources. Considerable progress has been made on observations and refinement of models to help explain low ozone loss at the mid-altitudes, the increase in UV dosage, the appearance of water vapor in the stratosphere, and possibly, of climate changes 50 million years ago.

The future in advanced molecular modeling offers the opportunity to solve grand challenge problems in environmental science:

- Fundamental advances in theory and computation will radically change the way we do science. Simulation science will become even more multidisciplinary. Simulation and computation will fully come of age as the third branch of science, fulfilling the promise of the past 20 years. Simulation will be key to coupling multiple temporal and spatial scales while maintaining accuracy. New models will emerge that will completely replace the techniques that have been used so far. For example, new, fast methods will replace 50 years of traditional quantum chemistry approaches and we will have new solvation models.
- The challenges will involve quantitative ab initio prediction of molecular-level chemistry of thermodynamics and kinetics with no empirical scaling, bridging the gap from the molecular scale to the microscopic (nano- and biological)

length and time scales. The current deficiencies in theory, computers, and software will require computational chemists to develop radical new approaches.

APPROACHES TO SOLUTIONS

Pollution Prevention: Green Process Technology

Green chemistry focuses on the design, at the molecular level, of manufacturing processes and products that are environmentally benign—reducing or eliminating the use of hazardous materials. A common definition of green chemistry is "the design, development and implementation of chemical processes and products to reduce or eliminate substances hazardous to human health and the environment."[22] Guided by a set of 12 principles,[23] green chemistry offers the potential to develop technologies that could provide an important new approach to environmental protection through pollution prevention.

We believe that the low viscosity of CO_2, coupled with its excellent wetting properties, will enable whole new classes of thin-film coating operations that will at the same time be environmentally responsible. These are likely to be important, not just for microelectronics applications but also for biomedical and nanotechnology formulations. Even though there are still many technical and economic barriers to the total acceptance of these technologies, we believe that environmental pressures as well as technical requirements for pure component systems with high uniformity will over time help "dry" CO_2-based processes play an increasingly important role in industrial environments.

[Ruben Carbonell, Appendix D]

As shown in Figure 3-2, green process technologies build on the input from multiple scientific disciplines. Many of the breakthroughs in green chemistry take place at the interfaces among these disciplines. Moreover, the contributions are not limited to the traditional interactions in which chemists, chemical engineers, physicists, and analytical chemists work together to develop a new process. The necessary collaboration may involve biologists, molecular biologists, and computational scientists as well.

For the chemical industry to thrive in the United States we will need im-

[22]Anastas, P. T.; Warner, J. *Green Chemistry Theory and Practice*, Oxford University Press, Oxford, UK, 1998.
[23]Poliakoff, M.; Fitzpatrick, J. M.; Farren, T. R.; Anastas, P. T. *Science* **2002**, *297*, 807-810.

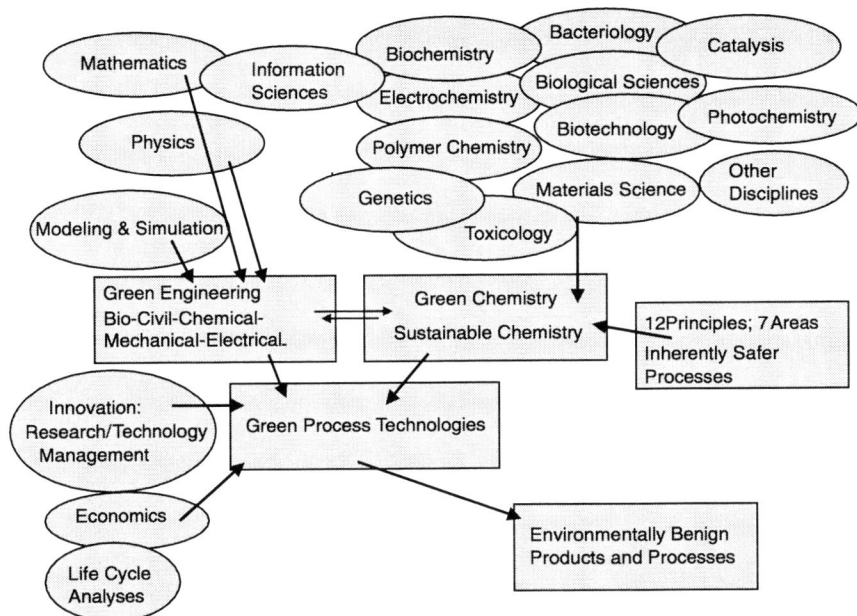

FIGURE 3-2 Sources of green process technologies. Influence diagram showing the information flow from scientific disciplines to green chemistry and engineering and then to green chemistry process technology and then environmentally benign products and processes.

proved efficiency and better ways of meeting regulatory hurdles. Otherwise economic decisions will be made to move our manufacturing offshore—which may not be a desirable outcome. One approach is to invest in better processes, and in many cases, in green process technologies. This is beginning to happen, with examples such as those described in Chapter 2.

A recent report from the RAND Science and Technology Policy Institute lists four major barriers to the development and implementation of new green technologies.[24] Finding ways to overcome these barriers will be a significant challenge to chemists and chemical engineers as they pursue their R&D agenda in the environmental arena:

• Need for additional research, technology development, or process engineering;

_____

[24]Lempert, R. J.; Norling, P.; Pernin, C.; Resetar, S.; Mahnovski, S. *Next Generation Environmental Technologies: Benefits and Barriers,* RAND, Arlington, VA, 2003; *http://www.rand.org/publications/MR/MR1682/.*

- Need to surmount infrastructure and integration barriers;
- Need to make the up-front investment; and
- Regulatory barriers.

The RAND report provides detailed analyses for a variety of case studies on next-generation technologies that demonstrate significant contributions by chemists and chemical engineers. Examples include

- Water purification: development of technologies such as new chemical methods, membrane technology, and ultraviolet irradiation could greatly reduce the quantities of chlorine from present levels;
- Liquid and supercritical CO_2 as reaction solvent: development of new processes could reduce the use of halogenated and other organic solvents;
- Depolymerization of polymers to monomers: conversion of polymers to the corresponding monomers can provide an alternative to recycling of the polymer, reduce landfill burden, and provide a new source of monomer with lower consumption of new raw materials;
- Biobased processes: the use of renewable feedstocks and biocatalytic processes reduce waste and greenhouse gas emissions while providing greater energy efficiency;
- New routes to hydrogen peroxide: new methods for direct synthesis of hydrogen peroxide (from hydrogen and oxygen) in a controlled, safe manner could provide a lower cost oxidant that reduces the use of chlorine. For example, in situ generation of hydrogen peroxide can be used to produce propylene oxide in place of the chlorohydrin route; and
- Dimethyl carbonate: new methods for synthesis and use of dimethyl carbonate could greatly reduce the use of highly toxic feedstocks such as phosgene; other waste streams (such as HCl) would be reduced as well.

Remediation

Soil and groundwater contaminated with hazardous materials create special challenges for chemists and chemical engineers. Determination of the composition and mobility of the contaminants, and the risks they pose to humans and the environment, often requires specialized analytical techniques. In some cases the hazardous nature of the contaminants may be reduced by natural attenuation due to chemical or biological activity in the soil, and a better understanding of the mechanism of attenuation can help to predict or accelerate the rate of hazard reduction. When remediation of the site is deemed necessary, cleanup or containment procedures must be tailored to the specific characteristics of the site.

The nation has a contamination legacy that results from practices by both the government and the private sector. Waste chemicals were dumped into trenches and waterways, contaminating hundreds of millions of tons of soil and water. It will also be necessary to dispose of the enormous radioisotope burden in tank wastes in the DOE complex. DOE is spending billions of dollars annually on cleanup of contaminated nuclear sites, but current methodologies may not be adequate to complete the task, even over a 50-year time span.[25] Extensive research efforts in biogeochemistry and reactive transport will be needed, based on the best ideas of geologists, chemists, chemical engineers, and microbiologists.

Catalysis and catalytic processes account for nearly 20% of the U.S. gross domestic product and nearly 20% of all industrial products. Chemical transformations in industry take a cheap feedstock (usually some type of hydrocarbon) and convert it into a higher-value product by rearranging the carbon atoms and adding functional groups to the compound. About 5 quads per year are used in the production of the top 50 chemicals in the United States and catalytic routes account for the production of 30 of these chemicals, consuming 3 quads. Improved catalysts can increase efficiency leading to reduced energy requirements, while increasing product selectivity and concomitantly decreasing wastes and emissions. A process yield improvement of only 10% would save 0.23 quad per year! In addition, production of the top 50 chemicals leads to almost 21 billion pounds of CO_2 emitted to the atmosphere per year. Improved catalysts can help reduce this carbon burden on the atmosphere. As new products become ever more sophisticated, the need to quickly develop new catalysts grows rapidly in importance. A fundamental understanding of chemical transformations is needed to enable scientists to address the grand challenge of the precise control of molecular processes by using catalysts.

[David Dixon, Appendix D]

Factors that must be considered in developing a remediation strategy include the chemical nature, quantity, and location of the contaminants; the permeability of the soil and how soil interacts with contaminants; and how various cleanup or containment methods may impact workers, the community, and remediation costs.

For shallow sites, it may be preferred to remove the contaminated soil and incinerate it or wash it ex situ. For deeper sites with porous soils it may be possible to flush out the contaminants with surfactants or solvents and treat the hazardous materials at the surface. If the contaminants are volatile, it may be possible to heat the soil and/or pump air or steam into the soil and capture the vaporized chemicals at the surface. In some cases, treatment chemicals may be

---

[25]See D. Dixon in Appendix D.

injected into the soil to react with the contaminants and produce a nonhazardous product. In other cases it may be preferable to immobilize the contaminants in situ by injecting a material that tightly binds to them or by heating the soil to form a virtually impermeable glass. In some cases an impermeable cap may be put over the contaminated site to mitigate the problem.

Innovative methods are under development to reduce remediation costs and to deal with particularly difficult sites. For example, electro-osmosis is being developed to move and treat contaminants in deep, low-porosity soils. Bioremediation is being developed to treat soils with microbes, and phytoremediation is being developed to utilize plants to remediate surface contamination. Novel chemical treatments are being developed to convert difficult contaminants, such as chlorinated hydrocarbons, to benign by-products. Permeable reaction barriers are being developed to trap or convert chemicals that pass through the barrier via natural migration or by electric field or pressure gradients. Improved geotextiles, landfill liners, and materials to contain radioactive wastes are being developed to enhance waste containment effectiveness.

Many challenges still remain to cost-effectively treat contaminants such as radioactive materials, and inert, tightly bound chemicals such as PCBs. The chemical community will play a major role in developing solutions to these and other complex contamination problems. In some cases—as with radioactive materials—it may not be feasible to destroy the waste, so long-term storage must be considered as an alternative. For such situations, chemists and chemical engineers will play an important role in developing safe and reliable approaches to containment of the waste (e.g., with improved materials for containers or in situ barriers that could limit migration of pollutants).

Cost-effectiveness

A major challenge to the chemists and chemical engineers in developing solutions to environmental problems is that of cost. Any solution that is proposed for a problem ultimately must be both technically and economically feasible. If cost precludes its implementation, then it is not an actual solution. Regulations sometimes require action that is accompanied by increased cost, but voluntary implementation of major change is unlikely without an accompanying economic advantage.

INTERFACES AND INFRASTRUCTURE

Many examples of collaborative work were discussed up during the workshop (Appendix G). These efforts have made substantial contributions to the development of environmental science and to improvements in the environment. In many ways environmental studies are inherently multidisciplinary as illustrated by Table G-2 in Appendix G. Finding ways to facilitate and enable such cross-disciplinary work will constitute a significant challenge, but it is one worth addressing because chemistry and chemical engineering will continue to contribute to fully understanding and solving environmental problems.

Workshop participants enumerated a variety of problems that currently inhibit effective collaborations, ranging from difficulty in communication across disciplines and the need for more cross-disciplinary educational programs to administrative barriers that inhibit research teams. Interdisciplinary activity in the academic environment would be enhanced by a reward structure that better recognizes the value of such collaborative efforts.

Enhancements in the infrastructure for education and research will be essential to future environmental progress. There was considerable sentiment among workshop participants that the current disciplinary structure of academic departments and funding agencies inhibits advances in the environmental arena. Communication and collaboration among federal funding agencies will also contribute to future progress. Research investment will need to address interdisciplinary activities, the need for development of new instruments, improved computational capabilities, and shared user facilities that may be too expensive for individual institutions.

4

Conclusions and Key Opportunities

The Challenges for the Chemical Sciences in the 21st Century Workshop on the Environment brought together chemical scientists and engineers from academia, government, national laboratories, and industrial laboratories, who provided a broad range of experience and perspective. Their discussions and presentations identified a wide variety of opportunities and challenges in chemistry and chemical engineering. These are documented throughout this report (see Appendix D and Appendix G for specific examples) and led the committee to its overarching conclusions:

Conclusion: Chemistry and chemical engineering have made major contributions to solving environmental problems.

Specific areas of accomplishment include

- major increases in analytical capabilities—detection, monitoring, and measurement;
- increased understanding of biogeochemical processes and cycles;
- advances in industrial ecology—new attitudes about pollution prevention;
- development of environmentally benign materials (e.g., CFC replacements);
- new methods for waste treatment and pollution prevention;
- green chemistry and new chemical processes;
- discovery of environmental problems and identification of their underlying causes and mechanisms; and
- development of improved modeling and simulation techniques.

Conclusion: Collaboration of chemists and chemical engineers with scientists and engineers in other disciplines has led to important discoveries.

These contributions have enhanced both basic understanding and the solution of environmental problems through work at the interfaces of the chemical sciences with biology, physics, engineering, materials science, mathematics, computer science, atmospheric science, meteorology, and geology.

Conclusion: Manifold challenges and opportunities in chemistry and chemical engineering exist at the interface with the environmental sciences.

By responding to these opportunities and challenges, the chemical sciences community will be able to make substantial contributions to

- fundamental understanding of the environment,
- remediation of environmental problems that currently exist,
- prevention of environmental problems in the future, and
- protection of the environment.

The stakes for responding to these challenges are high because regulatory decisions might cost or preserve billions of dollars, impact millions of human lives, or even determine the fate of entire species.

Much of the discussion at the workshop emphasized the interrelated nature of the many parts of the environment. Typically, it is not possible to take action in one area without creating at least the possibility of impacting other areas as well. In order to avoid such undesired consequences, a systems approach will be needed for the discovery and management of problems of the atmosphere, water, and soil. This will be necessary not only for understanding the complexity of each medium but for avoiding regulatory-driven tendencies to simply shift impacts from one medium to another.

A life-cycle systems approach, similar to what has been developed to evaluate energy impacts, will facilitate sound management of environmental impacts. This will provide a clear understanding of both where and when environmental impacts occur in the life of a product, process, or service. It also will make it possible to appreciate all impacts, and to see how interactions and alternatives at each point in a life cycle can influence other parts of the life cycle. For chemical processing and manufacturing, significant impact can occur at various stages, including extraction and preparation of raw materials, conversion of raw materials into products, separation and purification of materials, product distribution, end use of products, and final disposition after the useful life of products.

Conclusion: A systems approach is essential for solution of environmental problems.

The systems approach will be needed in several areas, including

- actions that affect any of the three principal environmental sinks (air, water, and soil) and the biological systems with which they interact, where attempts to manage each of them separately will surely transfer impacts from one medium to another;
- spatial management of environmental impact sources—where the impacts are generated in a processing and manufacturing sequence; and
- temporal management of environmental impact sources—when the impacts are generated in a processing and manufacturing sequence.

The use of systems approaches will necessitate simulation and modeling of enormously large and complex systems. This will require significant computational resources, intensive efforts in complex optimization, and formulation of mathematical models. Input and expertise from a broad array of scientific, engineering, and social disciplines will be an essential part of developing the necessary tools.

Conclusion: Solving environmental problems will require intensive mathematical modeling, complex optimization, and computational resources.

Systems approaches will necessitate extensive collaborations among a wide range of scientific, engineering, and social disciplines.

Workshop participants identified a broad array of research challenges, both in areas of fundamental understanding and for specific environmental problems.

Conclusion: Important opportunities exist for chemists and chemical engineers to contribute to a better understanding of the environment.

Many of these research opportunities will involve work at the interfaces with other disciplines or interdisciplinary collaborations with scientists and engineers from those disciplines. Just as these collaborations have led to significant progress in the past, they should be expected to play an important role in future efforts to fully understand and solve environmental problems. Examples include the need to understand (or better understand)

- structure-toxicity relations;
- chemical processes at the molecular level;

- biological and physicochemical interactions in response to environmental stresses;
  - fate and transport of anthropogenic chemicals;
  - biogeochemical cycles;
  - gas-to-particle conversion in the atmosphere;
  - functional genomics and the chemical processes that govern organism-environment relationships; and
  - chemical-gene interactions in the real environment.

As we continue to better understand the underlying science of the environment, further advances will require new tools and instruments.

Conclusion: Chemists and chemical engineers will need to develop new analytical instruments and tools.

These tools and instruments will have to function effectively in an increasingly complex research arena that involves measurements of vanishingly small quantities of substances in the presence of contamination from other chemicals, under circumstances that make sample acquisition difficult. They will have to address three principal areas of measurement:

1. laboratory analyses
2. field measurements
3. theoretical tools for modeling and comparison with experiment

Conclusion: Improved methods for sampling and monitoring must be developed.

Chemists and chemical engineers will have to address the challenges of sampling and monitoring—air, water, and soil—more extensively and more frequently than can be done now. This will require improvements in instrumentation, in sampling methodology, and in techniques for remote measurements.

Conclusion: The new approaches of green chemistry and sustainable chemistry offer the potential for developing chemical and manufacturing processes that are environmentally beneficial.

We are still in the early stages, but successful examples already have been reported. If the necessary investment is made in these new directions, chemists and chemical engineers will be able to make major strides in improving environmental quality.

Conclusion: Strong and continued support of the chemical sciences will be an essential part of the federal research investment for understanding, improving, and protecting the environment.

Chemists and chemical engineers will be able to respond effectively to the challenges described here only if they have the resources needed to carry out the necessary research. This impact of support will be enhanced if it facilitates interdisciplinary research and encourages industrial partnerships. The scientific progress resulting from such support will inform and enable the policy-making and decision process that is essential to future environmental improvement.

Appendixes

A

Statement of Task

The Workshop on The Environment is one of six workshops held as part of "Challenges for the Chemical Sciences in the 21st Century." The workshop topics reflect areas of societal need—materials and manufacturing, energy and transportation, national security and homeland defense, health and medicine, information and communications, and environment. The charge for each workshop was to address the four themes of discovery, interfaces, challenges, and infrastructure as they relate to the workshop topic:

- Discovery—major discoveries or advances in the chemical sciences during the last several decades
- Interfaces—interfaces that exist between chemistry–chemical engineering and such areas as biology, environmental science, materials science, medicine, and physics
- Challenges—the grand challenges that exist in the chemical sciences today
- Infrastructure—infrastructure that will be required to allow the potential of future advances in the chemical sciences to be realized

B

Biographies of the Organizing Committee Members

Mario J. Molina (Co-Chair) is Institute Professor at the Massachusetts Institute of Technology (MIT). He holds a chemical engineering degree from the Universidad Nacional Autonoma de Mexico; a postgraduate degree from the University of Freiburg, Germany; and a Ph.D. in physical chemistry from the University of California, Berkeley. He joined MIT in 1989 with appointments in both the Department of Earth, Atmospheric, and Planetary Sciences and the Department of Chemistry and was named MIT Institute Professor in 1997. Prior to joining MIT, he held teaching and research positions at the Universidad Nacional Autonoma de Mexico; the University of California, Irvine; and the Jet Propulsion Laboratory at the California Institute of Technology. He is a member of the Pontifical Academy of Sciences. He has served on the U.S. President's Committee of Advisors in Science and Technology, the Secretary of Energy's Advisory Board, the National Research Council (NRC) Board on Environmental Studies and Toxicology, and boards of the U.S.-Mexico Foundation of Science and other nonprofit environmental organizations. He is a member of the National Academy of Sciences and the Institute of Medicine. He has received several awards for his scientific work including the 1995 Nobel Prize in Chemistry.

John H. Seinfeld (Co-Chair) is the Louis E. Nohl Professor in the Divisions of Chemistry and Chemical Engineering and Engineering and Applied Science at the California Institute of Technology. He is a graduate of the University of Rochester, where he received a B.S. degree in chemical engineering, and Princeton University, where he received a Ph.D. in chemical engineering. In 1967, he joined the faculty of the California Institute of Technology. Through both experimental and theoretical studies, Seinfeld has made numerous contributions to our knowledge of the chemistry of the urban atmosphere; the formation, growth, and dynamics of atmospheric aerosols; and the role of aerosols in climate. His founding work in the field of mathematical modeling of the atmosphere eventually became written into the U.S. Clean Air Act. Seinfeld has received numerous honors and awards including the American Chemical Society Award for Creative Advances in Environmental Science and Technology, the 2001 Nevada Medal, and the Fuchs Award of the International Aerosol Research Assembly in 1998. He is a fellow of the American Association for the Advancement of Science and the American Academy of Arts and Sciences, and was president of the American Association for Aerosol Research. He was chairman of the NRC Panel on Tropospheric Ozone and the NRC Panel on Aerosols and Climate. Seinfeld is the author of more than 400 scientific papers and several books. He is a member of the National Academy of Engineering.

Mark A. Barteau (Steering Committee Liason) is Robert L. Pigford Professor and Chair of the Department of Chemical Engineering at the University of Delaware. He received his B.S. degree from Washington University in 1976 and his M.S. (1977) and Ph.D. (1981) from Stanford University. His research area is chemical engineering with specialized interests in application of surface techniques to reactions on nonmetals, hydrocarbon and oxygenate chemistry on metals and metal oxides, scanning probe microscopies, and catalysis by metal oxides.

Philip H. Brodsky retired from Pharmacia in 2002 as vice president responsible for corporate research and environmental technology, a position he held at Monsanto prior to its merger with Pharmacia and Upjohn. He received a Ph.D. in chemical engineering from Cornell University and has held various positions in research and research management at Monsanto. He serves as chair of the American Chemical Society's Committee on Chemistry and Public Affairs and on the industrial advisory boards of the Department of Chemical and Environmental Engineering at the University of Arizona and the Department of Chemical Engineering at Washington University. He has served on numerous advisory and review committees for the Department of Defense, Department of Energy, Environmental Protection Agency, and NRC and was a member of the boards of directors of the Industrial Research Institute, Inroads St. Louis, and MetaPhore Pharmaceuticals.

A. Welford Castleman, Jr. (BCST Liaison) is Evan Pugh Professor of Chemistry and Physics and Eberly Distinguished Chair in Science at the Pennsylvania State University and holds a joint appointment as professor in the Department of Physics. He has been a member and on the Advisory Board for the Particulate Materials Center at the Pennsylvania State University, currently serves in that capacity for the Consortium for Nanostructured Materials (VCU), and is a member of the Penn State Center for Materials Physics. He received a B.Ch.E. from Rensselaer Polytechnic Institute and his Ph.D. degree at the Polytechnic Institute of New York. He has been on the staff of the Brookhaven National Laboratory (1958-1975), adjunct professor in the Departments of Mechanics and Earth and Space Sciences, State University of New York, Stony Brook (1973-1975), and professor of chemistry and fellow of the Cooperative Institute for

Research in Environmental Sciences, University of Colorado, Boulder (1975-1982). Castleman is a fellow of the American Academy of Arts and Sciences and was a Fulbright senior scholar in 1989. He received the 1988 American Chemical Society Award for Creative Advances in Environmental Science and Technology and was awarded a Doktors Honoris Causa from the University of Innsbruck, Austria, in 1987. He is a member of the National Academy of Sciences.

Joseph M. DeSimone (BCST Liaison) is William R. Kenan, Jr., Distinguished Professor of Chemistry and Chemical Engineering at North Carolina State University and the University of North Carolina. He is also director of the National Science Foundation Science and Technology Center for Environmentally Responsible Solvents and Processes. He received his B.S. in chemistry form Ursinus College and his Ph.D. in chemistry from Virginia Polytechnic Institute and State University. His areas of interest include polymer synthesis in supercritical fluids, surfactant design for applications in interfacial chemistry, and polymer synthesis and processing—from fundamental aspects of chemical systems to the most efficient and environmentally friendly ways to manufacture polymers and polymer products.

Jean H. Futrell is Senior Battelle Fellow and Chief Science Officer at Pacific Northwest National Laboratory. Previously he was Willis F. Harrington Professor of Chemistry and Biochemistry at the University of Delaware. He received a B.S. from Louisiana Polytechnic Institute and a Ph.D. from University of California at Berkeley. Futrell's research program focuses on the application of reaction dynamics methods—particularly the use of crossed molecular beams—to investigate the detailed mechanism of ion activation in tandem mass spectrometry. He has served on the NRC's Chemical Sciences Roundtable and was chair of the Council for Chemical Research in 1999.

Parry M. Norling is American Association for the Advancement of Science (AAAS) Fellow at the RAND corporation. He retired in December 1998 after 33 years with the DuPont Company, where he held a number of R&D and manufacturing management positions and spent two years as corporate director of health and safety. From 1999 to 2001 he served part-time as Corporate Technology Adviser at DuPont supporting the chief science and technology officer. He is chairman of the Union of Pure and Applied Chemistry's CHEMRAWN (CHEMical Research Applied to World Needs) committee and a member of the IUPAC Bureau; he was chairman of the Industrial Research Institute (IRI) 1999-2000 and is currently a member of the Board of Directors of the American Creativity Association. He received an A.B. from Harvard University in physical sciences and a Ph.D. from Princeton University in polymer chemistry. His fields of expertise include R&D management, the functioning of human networks or communities of practices, improving the quality and effectiveness of innovation processes, assessing environmental technologies for sustainable development, understanding near-term nanotechnologies, and developing icephobic coatings.

Jeffrey J. Siirola (Steering Committee Liason) is a research fellow in the

Chemical Process Research Laboratory at Eastman Chemical Company in Kingsport, Tennessee. He received his B.S. in chemical engineering from the University of Utah in 1967 and his Ph.D. in chemical engineering from the University of Wisconsin-Madison in 1970. His research centers on chemical processing, including chemical process synthesis, computer-aided conceptual process engineering, engineering design theory and methodology, chemical technology assessment, resource conservation and recovery, artificial intelligence, nonnumeric (symbolic) computer programming, and chemical engineering education. He is a member of the National Academy of Engineering.

Christine S. Sloane is director of FreedomCAR and Advanced Technology Strategy at General Motors (GM) Corporation where she is responsible for technical strategy in advanced technology development programs and for the development and demonstration of technologies for energy efficiency and reduced emissions under the FreedomCAR program. Prior to 2002, Sloane served as director, Technology Strategy for Advanced Technology Vehicles, focused on electric drive and hybrid drive systems. Earlier she served as director, Environmental Policy and Programs, responsible for global climate issues and for mobile emission issues involving advanced technology vehicles. From 1994 to 1999, Sloane served as chief technologist for the development and demonstration team for Precept, GM's 80 mile-per-gallon 5-passenger demonstration vehicle. She also served as GM's technical director for the Partnership for a New Generation of Vehicles (PNGV) where she was responsible for guiding and implementing the development of energy conversion and materials technologies through research and development at national laboratories, universities, and automotive suppliers. Her earlier research interests included aerosol chemistry and physics, air quality and visibility, manufacturing and vehicle emissions, and environmental policy. Sloane has authored more than 80 technical papers and coedited one book. She has served as department head of atmospheric sciences at Battelle Pacific Northwest Laboratories. She received her Ph.D. from MIT in chemical physics.

Isiah M. Warner is Boyd Professor and Philip W. West Professor of Analytical and Environmental Chemistry at Louisiana State University (LSU). He received his B.S. from Southern University in 1968 and worked as a research chemist with Battelle Northwest for five years before receiving his Ph.D. from the University of Washington in 1977. He served on the faculties of Texas A&M University and Emory University before joining LSU in 1992. Warner's research focuses on the areas of molecular spectroscopy and separation science. He has published more than 200 peer-reviewed manuscripts and several book chapters, and has coedited two books. He has won numerous awards for his research, teaching, and mentoring, including the year 2000 Eastern Analytical Symposium Award for Outstanding Achievements in the Fields of Analytical Chemistry; the 1997 Presidential Award for Excellence in Science, Mathematics, and Engineering Mentoring; the year 2000 AAAS Lifetime Mentor Award; and the year 2000 Council for Advancement and Support of Education (CASE) Louisiana Professor of the Year Award from the Carnegie Foundation.

C

Workshop Agenda

Agenda
Workshop on the Environment
Challenges for the Chemical Sciences in the 21st Century

National Academy of Sciences
Arnold and Mabel Beckman Center
Irvine, California

SUNDAY, NOVEMBER 17

   7:30   Breakfast and Registration

SESSION 1: THE CHEMISTRY OF THE ENVIRONMENT
   8:10   Introductory remarks by organizers—background of project
   8:15   **DOUGLAS J. RABER,** *National Research Council*
   8:20   **MATTHEW V. TIRRELL,** Co-Chair, Steering Committee, *Challenges for the Chemical Sciences in the 21st Century*
   8:25   **MARIO J. MOLINA, JOHN H. SEINFELD,** Co-Chairs, Organizing Committee, *Workshop on the Environment*
   8:30   **BARRY DELLINGER,** *Louisiana State University*
               Origin and Control of Toxic Combustion By-Products
   9:05   DISCUSSION
   9:25   **CHARLES E. KOLB,** *Aerodyne Research, Inc.*
               Measurement Challenges and Strategies in Atmospheric and Environmental Chemistry
 10:00   DISCUSSION

 10:20   **BREAK**

10:50   JANET G. HERING, *California Institute of Technology*
Biogeochemical Controls on the Occurrence and Mobility of Trace
Metals in Groundwater
11:25   DISCUSSION
11:45   Lunch

Session 2: The Chemistry of the Environment (Part 2)
1:00   MARK THIEMENS, *University of California, San Diego*
Measurements of Stable Isotopes in Atmospheric Species
1:30   DISCUSSION
1:50   FRANÇOIS M. M. MOREL, *Princeton University*
Chemistry of Trace Elements in Natural Waters
2:20   DISCUSSION

2:40   BREAKOUT SESSION: **DISCOVERY**
*What major discoveries or advances related to the environment
have been made in the chemical sciences during the last several
decades?*

3:45   **BREAK**
4:00   Reports from breakout sessions (and discussion)
5:00   RECEPTION
6:00   BANQUET
Speaker: WILLIAM F. FARLAND, Acting Deputy Assistant
Administrator for Science, *U.S. Environmental Protection Agency*

MONDAY, NOVEMBER 18

7:30   Breakfast

Session 3: Manufacturing and Green Chemistry
8:00   UMA CHOWDHRY, *DuPont*
Sustainable Growth in the Chemical Industry
8:30   DISCUSSION
8:50   RUBEN G. CARBONELL, *North Carolina State University*
CO_2-Based Technologies
9:20   DISCUSSION

9:40   BREAKOUT SESSION: **INTERFACES**
         What are the major environment-related discoveries and
         challenges at the interfaces between chemistry–chemical
         engineering and other disciplines, including biology, information
         science, materials science, and physics?

10:45   **BREAK**
11:00   Reports from breakout sessions (and discussion)
12:00   Lunch

Session 4: Manufacturing and Green Chemistry (Part 2)

1:00   THOMAS W. ASMUS, *DaimlerChrysler Corporation*
         Diesel Engines for a Clean Car?
1:30   DISCUSSION
1:50   MICHAEL K. STERN, *Monsanto Company*
         Environmentally Sound Agricultural Chemistry: From Process
         Technology to Biotechnology
2:20   DISCUSSION

2:40   BREAKOUT SESSION: **CHALLENGES**
         What are the environment-related grand challenges in the
         chemical sciences and engineering?

3:45   **BREAK**
4:00   Reports from breakout sessions and discussion
5:00   ADJOURN FOR DAY

TUESDAY, NOVEMBER 19

7:30   Breakfast

Session 5: Environmental Remediation and Modeling

8:00   JAMES G. ANDERSON, *Harvard University*
         A Developing Generation of Observation and Modeling Strategies
8:30   DISCUSSION
8:50   DAVID A. DIXON, *Pacific Northwest National Laboratory*
         Modeling and Simulation for Environmental Science
9:20   DISCUSSION

9:40   BREAKOUT SESSION: **INFRASTRUCTURE**
            What are the issues at the intersection of environmental studies
            and the chemical sciences for which there are structural
            challenges and opportunities—in teaching, research, equipment
            and instrumentation, facilities, and personnel?

10:45   **BREAK**
11:00   Reports from breakout sessions (and discussion)
12:00   Wrap-up and closing remarks
            MARIO J. MOLINA, JOHN H. SEINFELD, Co-Chairs, Workshop
            Organizing Committee
12:15   ADJOURN

D

Workshop Presentations

A DEVELOPING GENERATION OF OBSERVATION AND MODELING STRATEGIES

James G. Anderson
Harvard University

In the next few decades, several problems of considerable societal interest will emerge with a central theme of the balance between societal objectives and scientific curiosity-driven research, both of which are very important for the future. Three problems of key importance in the area of observations and modeling are:

- Migration of nitrate, sulfate, heavy metals, and organic soot from urban or regional areas into the broader environment and associated public health issues;
- Forecasting climate change and testing the forecast in a way that is acceptable to a much broader range of individuals involved both in science and in public policy; and
- Ultraviolet dosage, which is in many ways a statement of what it is that society really cares about, particularly in connection with the ozone question.

The way in which we attack these problems—the observations, the modeling, and the way public understanding evolves—are crucial issues. For example, Charles Kolb's presentation provides a beautiful introduction to the issue of undersampling and the requirement for significant advances in core technology that must underpin the scientific case leading to effective public policy.

BOX 1
Experimental Requirements

- High spatial resolution observation of fluxes, isotopes, reactive intermediates, tracers, particles, aerosol, radiation field, etc.
- Regional and national budgets of carbon, photochemical oxidants, particulates
- Region-specific studies:
  ○ US and its boundaries, inner tropics–Intertropical Convergence Zone (ITCZ)
  ○ Indian Ocean–Indian subcontinent, Asia
- Seasonal signature:
  ○ Winter-summer monsoons, wet-dry tropical
  ○ Winter-summer temperate and high latitudes
- Scientific objectives:
  ○ Vertical fluxes from biogenic sources and sinks
  ○ Urban and regional sources and sinks driven by human activity; gas-to-particle conversion mechanisms
  ○ Convective redistribution of water vapor, sulfates, nitrates, soot, carbon, heavy metals; free-radical catalysis and associated photochemical mechanisms
- Structure: extremely sharp gradients, localized fluxes, discontinuities across specific trajectories

---

The issue of carbon sources and sinks is strategically tied back to the question of nitrate, sulfate, heavy metal, and organic soot emission. If we can attack one problem and solve it, we will be attacking both. The question of how nitrate will affect human health is closely tied to carbon sources and sinks, which in turn are linked to climate. Consequently, answering the carbon-nitrate-sulfate question requires very high spatial resolution of fluxes, isotopes, and reactive intermediates.

There are particular problems with region-specific studies (Box 1). For example, the Indian subcontinent—as it links into the tropical region during the monsoon season—is as different from the other seasons as any two regions on Earth. In addition, the way in which these systems couple from the regional to the global scale are extremely important for the issue of prediction. Describing a system and understanding it well enough to predict really separate the strategies of observations and modeling.

In addition to seasonal signatures, we want to understand vertical fluxes and urban regional source sinks driven by human activity. We understand chemical transformations very well as illustrated by the Los Angeles basin. We know exactly what reactions are taking place, but the lack of specificity on the location and strength of sources and the way those couple into the chemical transformation serves to prevent a link between science and public policy on that question.

The strategy for analyzing NO_x across the globe by comparing satellite observations, models, and in situ aircraft observations hinges on the fact that the NO to NO_2 ratio goes up dramatically with increasing height in the troposphere. For example, NO_2 measurements from the European Space Agency's Global Ozone Monitoring Experiment (GOME) satellite provide a way of analyzing NO_2 in the boundary layer and just above the boundary layer. Therefore it possible to test the differences between the GOME satellite measurements and global chemical models. Those differences tell us where we have to attack the problem with a more sophisticated set of measurements. In particular, the large differences between the modeled and satellite-observed concentration fields establish the priorities for aircraft field campaigns that are the only means by which we can adjudicate these differences.

Now consider the third question—that of UV dosage. Here we have an emerging marriage between the atmospheric dynamics, chemistry, and medical communities because malignant melanoma, as we'll see in a moment, is a huge and increasing health issue. In fact, skin cancers are the only cancers worldwide that are increasing in a statistically important way in the face of developing medical methods. So it's this union between the dynamics and the loss of ozone at midlatitudes (Figure 1) that provides the fundamental information that most of the ozone loss is taking place in the very lower part of the stratosphere. This leads to several questions:

- Which mechanisms are responsible for the continuing erosion of ozone over mid latitudes of the Northern Hemisphere?
- Will rapid loss of ozone over the arctic in late winter worsen? Are these large losses coupled to midlatitudes?
- How will the catalytic loss of ozone respond to changes in boundary conditions on water and temperature forced by increasing CO_2, CH_4, and so forth?

How we respond in the future requires an understanding of the mechanisms that control the long-term erosion of ozone at midlatitudes. We know why ozone is destroyed over the Antarctic and over the arctic, but there is also a seasonally dependent midlatitude ozone loss. The months of March, April, and May define a key period—when schools let out, final exams are over, and the younger population gets a large episodic dose of ultraviolet radiation. This March-April-May period shows the largest long-term erosion over the past 20 years, approaching 10% ozone loss per decade. However, the strategy one takes depends, to an extent, on whether this is a societal or a scientific issue.

Let's take the position that we don't know the mechanisms that control ozone erosion over midlatitudes (which at this juncture is true) and assume that the erosion of ozone will continue as it has in the past decades. Then we're going to look at the coupling with the large ozone losses in the arctic triggered by conversion of inorganic chlorine to free-radical form on ice particles or cold liquid aero-

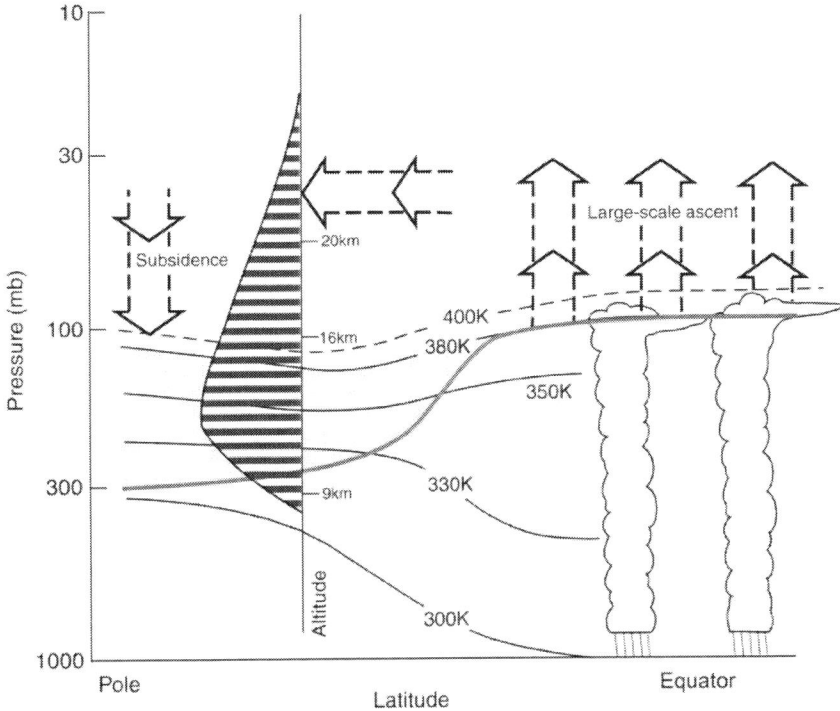

FIGURE 1 Midlatitude ozone loss.

sols each winter in the polar vortex. We come to the conclusion, in such an analysis, that it's the dynamical structure of the atmosphere that underlies the fundamental cause of this midlatitude ozone loss, not chemistry directly.

From the societal perspective, the raw numbers are large and important for basal cell carcinoma: 800,000 cases per year. Until we understand the mechanism of midlatitude ozone loss, a simple extrapolation may be inaccurate, but it provides an important reference for discussion (Figure 2). By simple extrapolation there would be an increase from 800,000 cases of basal cell carcinoma in the United States annually to nearly 1.9 million by 2060. The logarithmic dependence of the cross section results in a 2% increase in UV at these optical depths for a 1% decrease in ozone. The biological amplification factor emerges from the medical community, and it's a change in human morbidity. For malignant melanoma, the numbers are very much smaller, but the fractional death rate, as opposed to morbidity, is much greater.

Consider the pattern of ozone over the Northern Hemisphere, and ask if this simply results from large ozone concentrations over the arctic migrating back into midlatitudes. That would correspond to chlorine- and bromine-catalyzed

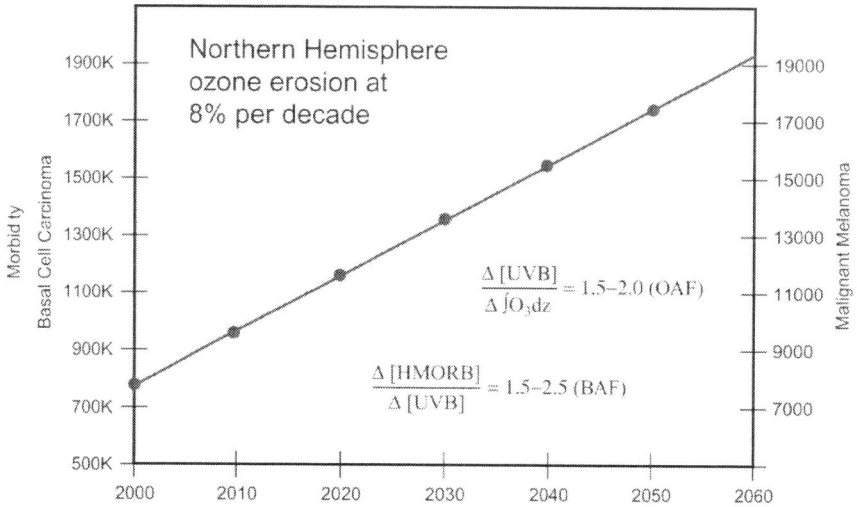

FIGURE 2 Hypothetical trend for basal cell carcinoma if the trends in midlatitude ozone erosion continue unchanged.

ozone destruction over the winter arctic merging back into midlatitudes. It's a completely reasonable, simple, understandable hypothesis, and it emerges from the fact that in the early 1970s we had this dome of ozone that represented the sequestration of ozone moving from the low latitudes into the high latitudes. As the ozone moves northward and downward, it's sequestered in this dome, and that's the way the world has worked for millions and millions of years. Until the late 1990s, as the ozone layer began to thin in many winters, it was emulating the Antarctic in a dramatic way. Does this low-ozone air created in the winter vortex flow back into midlatitudes, causing the observed minimum in the March-April-May period? An analysis of data from the last National Aeronautics and Space Administration (NASA) arctic mission indicates clearly that this does not occur.

There are indeed large ozone losses in the vortex, but all of the large-scale flow is from the tropics northward and downward. Also, because we know the seasonal phase of CO_2 and water over the tropical tropopause, we know that there is no communication backward from the polar regions to midlatitudes in these key months of ozone loss. Therefore, it isn't simply ozone-depleted arctic air moving back into midlatitudes.

Does long-term ozone erosion result from chemical loss of ozone at midlatitudes in the lower stratosphere? Susan Solomon made the very reasonable suggestion that the penetration of cirrus clouds and cold aerosols into the lower stratosphere initiates the conversion of inorganic chlorine to free-radical form, and she carried out a number of modeling studies that support this.[1] However, evidence garnered from hundreds of crossings of the tropopause by the ER-2 aircraft—which provides very high-resolution simultaneous measurements of tropopause position, water vapor and temperature, percentage of observations with ice saturation, and ClO concentrations—demonstrates that cirrus clouds and cold aerosols capable of providing the heterogeneous site for inorganic chlorine to free-radical conversion do not exist. The absence of observed ClO is the most compelling point. It would take about a 50-200 parts per trillion (ppt) of ClO to drive the ozone loss that's observed, but the experimental measurements showed less than 2-4 ppt of ClO, all the way up to 4 km above the tropopause.

Consequently, we do not believe that in situ loss of ozone is responsible for the long term trend in midlatitude ozone erosion, even though that would be the most reasonable explanation. This brings us back to the fundamental unsolved question of the coupling between the tropics and high latitude. As the temperature of the ocean surface warms in response to increasing CO_2 forcing, how does it affect the boundary condition on the entrance of water vapor into the stratosphere? The tropical tropopause constitutes a valve that desiccates the stratosphere and strictly controls water vapor entering the stratosphere from the troposphere.

Large-scale ascent operating above the tropical tropopause delivers this mixing ratio of water vapor into the high latitudes. For temperatures running from 192 to 200 K and a given water vapor curve of 6 parts per million (ppm), the trigger point for formation of high ClO is about 195 K. We have verified this experimentally using ER-2 observations with trajectory calculations demonstrating the temperature history of air parcels moving within the vortex.

Consequently, the amount of water vapor, as it increases in the stratosphere in response to increasing temperatures at the tropical tropopause, would instigate a shift in the threshold temperature required to instigate ClO formation to temperatures above 195 K, thereby aiding the dramatic loss of ozone in the arctic winter vortex. At the same time, the increase in water vapor in the vortex induces radiative cooling that drops the temperature of the wintertime lower stratosphere at high latitudes, thus exacerbating the destruction of ozone by chlorine radicals. The water vapor is the crucial quantity, and the structure of the tropics turns out to be the centerpiece for understanding all of these issues linking climate and trends in UV dosage.

What is needed is to couple the boundary layer in the tropics all the way up through the tropical transition layer into the stratosphere (Figure 3). The geographic coverage is crucial. The eastern tropical Pacific over Central America is dramatically different from the western tropical Pacific. It is the structure of con-

[1]Solomon, S.; Borrmann, S.; Garcia, R.R.; Portmann, R.; Thomason, L.; Poole, L.R.; Winker, D.; McCormick, M.P. *J. Geophys. Res.* **1997**, *102(D17),* 21411-21429.

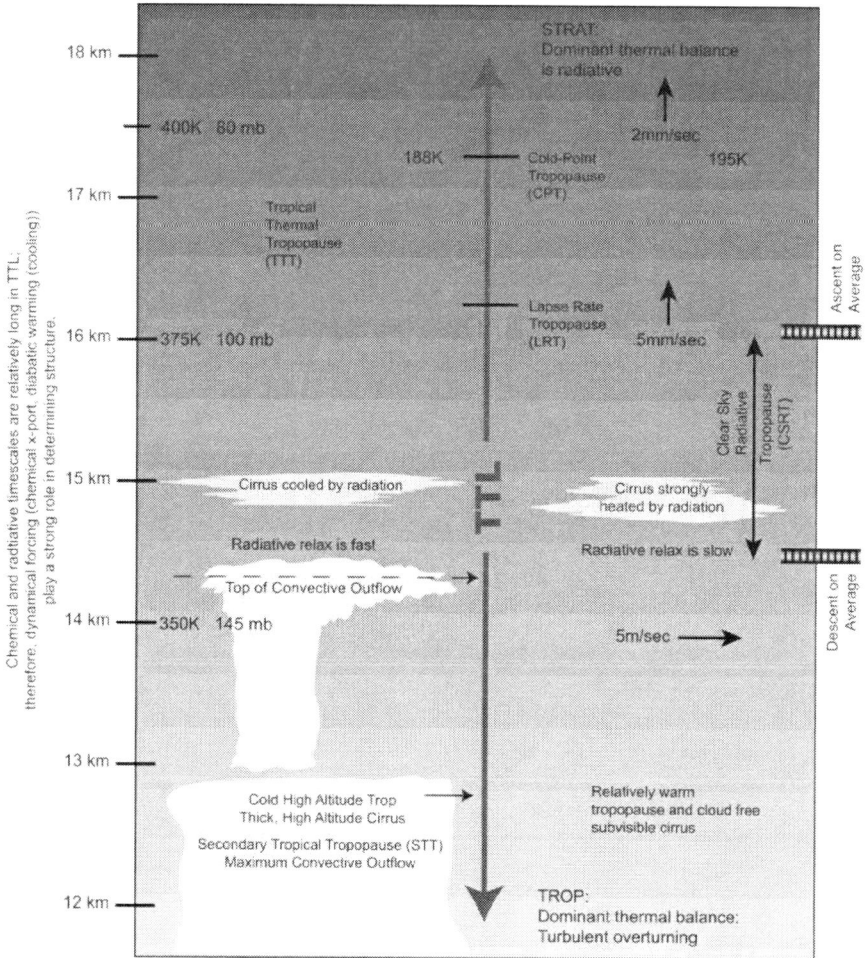

FIGURE 3. Deep convection in the tropics.

vection in the tropics, the control of water vapor in the middle-upper troposphere, and the formation of high-altitude cirrus in the region between 13- and 18-km altitude that must be understood before predictions of the impact of climate change can be made.

There are important suggestions in the paleorecord. Consider the Eocene 50 million years ago. In central Wyoming, there were turtles, alligators, and palm trees, a combination of plant and animal life that extended into central northern Canada. It was an era defined by deep ocean temperatures, running 10 K above present and warm polar sea surface temperatures. The Northern Hemisphere continental interiors were warm throughout the year. There was no glaciation. How this occurred is an important question that carries crucial messages for today— including how we attack the problem scientifically.

The mechanisms responsible for Eocene climate have been suggested to be enhanced meridional heat fluxes. The ocean is always involved here, with reorganization of the atmospheric circulation, enhanced greenhouse warming due to high carbon dioxide, and reductions in global topography. Yet none of the models have been able to capture the gentle difference in temperature between the tropics and high latitudes. This was explained by Sloan and Pollard in 1998, when they introduced polar stratospheric clouds into the model.[2] These are the site for heterogeneous reaction, but they also are profoundly important for trapping infrared radiation. They introduced methane in our favorite reaction to produce the oxidation leading to the formation of water. They proposed that the methane came from swamps and wetlands.

The Eocene went on uninterrupted for 10 million years, but the chemical lifetime of methane is about 7 years. Consequently, we don't believe that this is the explanation. We believe that polar stratospheric clouds are trapping high-latitude radiation in the infrared, but we believe the mechanism comes from the following: If CO_2 enters the system or heat moves northward, the gradient between the tropics and the poles begins to soften, leading to reduced excitation of gravity waves and planetary waves driving from the troposphere up into the stratosphere. As the flux of upwelling gravity waves and planetary-scale waves is reduced, so is the wave drag effect, which is the dominant pump that lifts material up in the tropics and pushes it down at high latitudes. If the effectiveness of that pump is reduced, the system relaxes back over the tropics so that the boundary condition on water vapor increases, allowing significantly more water to get into the system. We believe that this is the crucial climate state, and it is at the heart of our understanding of the current climate and also of UV dosage.

This brings us back to analyzing the meridional cross section in three dimensions from the equator to the pole. A long-duration balloon that could remain in the lower stratosphere would allow us to sample vertically by lowering a package. This is similar to stratospheric experiments from the 1980s, but the subtlety of the connection of the dynamics with the radiative and chemical properties demands an entirely different look.

We know that we can scan 10 km back and forth, but now we have some tremendous help from fuel cells. The technology of fuel cells allows us to use solar energy for balloon flights lasting several months. Solar energy drives a very efficient propeller to position the balloon at whatever latitude we want to scan. This energy technology has produced major breakthroughs. Even though other approaches are important, we think that the subtlety of this connected system will

[2]Sloan, L.C.; Pollard, D. *Geophys. Res. Lett.* **1998**, *25(18)*, 3517-3520.

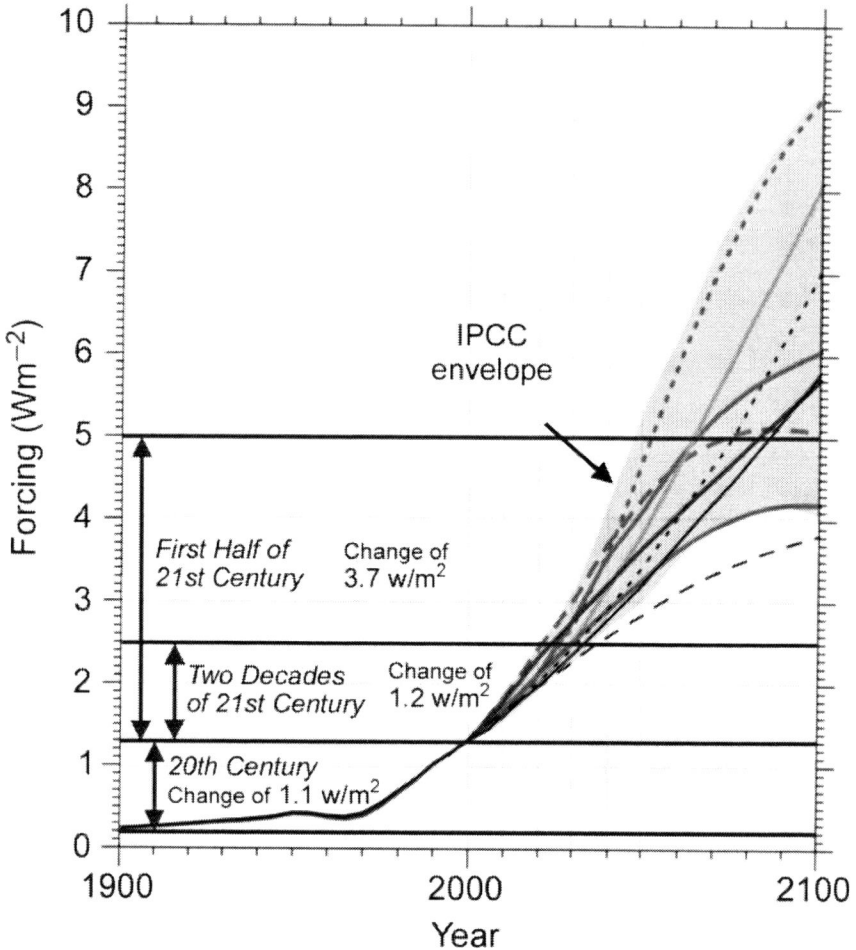

FIGURE 4 Atmospheric radiative forcing trends.

emerge only out of observations of tracers, particles, and the velocity components associated with this when done in a very sensitive way from an immobile platform.

We must still address the question of testing climate forecasts. Atmospheric CO_2 levels are rapidly increasing, as documented in the Intergovernmental Panel on Climate Change (IPCC) report.[3] A number of different scenarios have been

---

[3]Intergovernmental Panel on Climate Change (IPCC), *Climate Change 2001: The Scientific Basis,* Houghton, J.T.; Ding, Y.; Griggs, D.J.; Noguer, M.; van der Linden, P.J.; Xiaosu, D., Eds., Cambridge University Press: Cambridge, New York, 2001 (*http://www.grida.no/climate/ipcc_tar/*).

developed, and it is clear that we're hugging the upper boundary on the release of CO_2 that drives the forcing. We are projecting nearly 4 W/m^3 by the middle of the twenty-first century (Figure 4). This is emerging out of the natural variability, and it will be extremely important.

The final point is the question of societal objectives. If you look at the question of climate forecast from societal objectives, what we need is an operational forecast that is tested and trusted. Yet this nation has no operational climate forecast. We have no test of the veracity of that forecast; and until we do, we cannot deliver what is needed to the public.

The backbone of the climate forecast, of course, is the operational model that links the short-term El Niño scale to the longer term. The observing system is the key challenge for testing the veracity of calculations. Carbon sources and sinks have been discussed. I believe that upper ocean observations, climate data records at the surface, and benchmark observation that establish the long-term evolution of the climate in an absolute sense constitute the centerpiece of what must be done.

DIESEL ENGINES FOR CLEAN CARS?

Thomas W. Asmus
DaimlerChrysler Corporation

While the Diesel engine has been growing in popularity in the light-duty vehicle segments in many parts of the world, it has essentially stalled in the United States in these segments. In large part this results from the absence of financial incentives given the low levels of fuel taxation in the United States, plus an image problem based largely on experiences of two decades ago. Since that time, Diesel engine technology has improved in many ways, and this manifests itself as increased performance and fuel economy accompanied by reduced emissions, odor, and noise. In large part this has been made possible, not by any scientific breakthroughs, but rather by machine design and manufacturing technology advancements particularly in fuel-injection system components. In the light-duty vehicle segments today the typical fuel economy expectation of a Diesel engine is 40% greater than that of the gasoline engine at equal vehicle performance on a fuel volume-normalized basis. Because Diesel fuel has roughly 15% more energy per gallon than gasoline, the benefit is roughly 25% on an energy-normalized basis.

The U.S. regulatory mandates on criteria emissions have been within reach for gasoline and Diesel-powered vehicles to the present, but in 2007 the mandates will be beyond reach for the Diesel with any kind of sensible emissions abatement scheme. This, combined with a very weak business case based on low fuel taxation, effectively discourages U.S. industry investment in Diesel—the only practical high-fuel-economy alternative to the more conventional and well-known gasoline engine technology. While the cost of a modern, high speed Diesel engine is substantially more than its gasoline counterpart, it is the most cost attractive alternative to conventional gasoline.

Evolution of the Diesel Engine

Diesel engine combustion is a highly mixing-limited, stratified-charge process, and herein lies its formidable efficiency advantage over homogeneous-charge gasoline as well as its challenges with respect to NO_x and particulate matter (PM). Whereas the homogeneous-charge relies on inlet throttling for load control, the Diesel relies only on injected fuel quantity for this. Early attempts at managing this mixing-limited process often involved using an outside source of compressed air to assist the fuel atomization process. Later, much attention focused on manipulating in-cylinder air flows to hasten the mixing process. Meanwhile, machine design and manufacturing specialists found practical means to increase fuel-injection pressures and to better manage machining process for better control of fuel-injection precision. Ultimately there was less reliance on in-cylinder air flow manipulation to support the mixing process; hence higher engine speeds were enabled by faster combustion, while NO_x and PM emissions were reduced. With this, the pre-combustion chamber Diesel became obsolete, and direct injection (open combustion chamber design) became the standard for Diesel engines of most all sizes and duty cycles. With this came reductions in combustion chamber surface area and also in-cylinder turbulence, both of which contribute to reduced heat losses. Typically a 15% increase in thermal efficiency accompanied this shift. High-pressure common-rail fuel systems are state of the art, they provide substantially improved combustion manipulation ability compared to all progenitors, and turbocharging has become standard on all small high-speed, automotive Diesel engines. (Turbocharging Diesel engines is highly beneficial to vehicle fuel efficiency since smaller engines with lower friction can be used. Any fuel efficiency benefit derived from turbocharging gasoline engines is somewhat more conditional.) Relative to conventional gasoline engines, these produce significantly higher torque density and near-competitive power density. At least a portion of Europeans' enthusiasm for Diesel power is their highly desirable drivability and performance characteristics.

Engine-Out Emissions

The fuel efficiency advantage of the Diesel over conventional gasoline is based primarily on the means of load control via injected fuel quantity while the air flow is essentially unrestricted (i.e., there is no inlet throttling); therefore the gas exchange (or pumping) loss is minimal. At light loads, therefore, the overall air-fuel ratio is sufficiently high that homogeneous-charge flame propagation is not possible. Hence, stratified charge, diffusion-limited combustion is the principal heat release mechanism. Therefore, combustion and post-flame chemistry occur over a wide range of air-fuel ratios. The fraction of heat release that occurs in the range of stoichiometric yields very high temperatures and produces high levels of NO_x. The portion of heat release that occurs in fuel-rich zones (equivalence ratio = 2 or greater; temperatures = 1400 K or higher) tends to produce PM or soot. Manipulation of local air-fuel ratios is the goal of fuel-injection and in-cylinder air-motion strategies.

As described above, recent improvements in fuel-injection equipment have produced remarkable reductions in both of the principal emittants. This has been achieved mainly via higher injection pressures and precision control of the in-stantaneous rates of injection (with techniques described as rate shaping, split injections, pilot injection, post injection, etc.). These advanced fuel-injection tech-niques enable a measure of control over the local mixing processes and thus the local air-fuel ratios. Trends toward increased levels of premixed burning tend to be beneficial in terms of reducing both of the principal emittants. These injection-mixing strategies, while very desirable, are limited by the need for timing preci-sion of the thermal autoignition process (i.e., timing of thermal autoignition is compromised when temporal proximity between injection and ignition is in-creased). While the emissions reductions resulting from these advancements are impressive, attainment of the so-called Tier 2, bin 5 legislated limits to be en-forced in 2007 is not possible by these means alone.

At the far end of this spectrum is homogeneous-charge compression ignition (HCCI), which has been the subject of considerable research effort over the past several decades. This scheme strives to produce a truly (or nearly) homogeneous charge, which under light-load conditions, corresponds to air-fuel ratios too lean to support either flame propagation or diffusion-limited combustion. Hence, multisite thermal autoignition will occur, provided that a particular temperature threshold is attained. This process is capable of producing very low emissions, but timing control of the thermal autoignition process is severely compromised, since the injection event is no longer capable of precisely triggering heat release. Hence, this scheme is readily demonstrable on a laboratory scale but is generally incapable of sustained operation upon load changes due to perturbations in ther-mal equilibria.

Diesel Engine Aftertreatmen

Significant publicly and privately funded R&D activities have been in place in this area for the past decade, and while incremental improvements have been realized, they all fall short of what is necessary to achieve legislated requirements for 2007 with respect to both NO_x and PM. Complicating matters are low exhaust temperatures at light loads characteristic of automotive-type duty cycles and the presence of significant quantities of oxygen in the exhaust stream. In addition, chemical reduction of NO_x in a strongly oxidizing exhaust environment presents additional and formidable challenges. Obviously, chemical reduction of NO_x requires that reductant chemicals be added to the exhaust in a manner compatible with the particular aftertreatment scheme. The so-called urea-SCR (selective catalytic reduction) scheme is perhaps the most effective means of reducing tailpipe NO_x emissions. Aqueous urea is metered as precisely as possible and stoichiometrically with respect to NO_x into the exhaust just upstream of the SCR catalyst, the first stage of which hydrolyzes urea to ammonia. The final stage of the SCR system is an oxidation catalyst that oxidizes any excess of ammonia that may not have been consumed in the SCR catalyst itself. Despite functional attributes of the SCR approach, it comes with two formidable drawbacks: (1) a urea infrastructure would be needed, and (2) the U.S. Environmental Protection Agency (EPA) has virtually ruled out this option.

The so-called NO_x absorber (or trap) is generally seen as the most attractive option for the United States Although generally less effective than the urea SCR approach, this uses Diesel fuel as a source of chemical reductant. The catalyst is typically barium and platinum on an alumina substrate, where nitric oxide is catalytically oxidized to nitrogen dioxide and then reacts with the barium to form barium nitrate. When the barium becomes completely nitrated, the trap is regenerated by the addition of Diesel fuel somewhere upstream of the catalyst. The twofold role of the secondary fuel addition is (1) to consume all excess oxygen and (2) to produce cracked products (H_2, CO, and hydrocarbons in decreasing order of effectiveness) to effect chemical reduction of the trapped NO_x. NO_x absorbers are extremely sulfur sensitive and are rendered inactive when fuel-bound sulfur, which leaves the engine as SO_2, becomes catalytically oxidized to SO_3 whereupon it reacts "irreversibly" with barium to form the comparatively more stable barium sulfate. Desulfation of NO_x absorbers requires temperatures sufficiently high as to degrade the catalyst. Fuel sulfur levels will directly impact desulfation frequency and hence catalyst life expectancy. This along with the denitration frequency will determine the fuel economy penalty associated with this form of NO_x aftertreatment.

PM traps are typically ceramic wall-flow filters that trap PM rather effectively. PM trap regeneration, on the other hand, has proven to be rather more challenging. Typically an additional source of heat must be added to effect the initiation of oxidative regeneration. Once regeneration is initiated, the exothermicity of the process can (and often does) overheat the filter, causing irreversible damage. Various means may be employed to reduce the required regeneration temperature.

These basic aftertreatment elements have been demonstrated in a wide variety of different configurations. All are very sensitive to duty-cycle diversity, fuel quality, packageability, and control approaches. High-mileage durability has not been thoroughly established for any of these systems. Some demonstrations have been reported for limited duty-cycle testing typically with "tailored," open-loop regeneration control.

Overall Conclusions

To date there has been no demonstration of system compatibility with the legislated emissions limits to be enforced under Tier 2, bin 5 (70 mg NO_x per mile and 10 mg PM per mile) in 2007. (It is typically necessary to target 50% of these emissions limits to ensure high-mileage conformance.) In addition to the afore-mentioned technical obstacles, at current levels of fuel taxation in the United States and customer indifference to matters of fuel economy, private-sector in-vestment in Diesel system technologies is discouraged. In the automotive indus-try, lead times are such that without proven emissions capabilities as of this date, it is highly unlikely that this technology will be available in the United States by 2007 based on the world's most stringent emissions regulations. It is also note-worthy that Diesel technology is the most promising and cost-effective among the major fuel-economy enabling technologies.

Opportunities Through the Chemical Sciences

In the interest of mitigating some of the risks that could deprive the U.S. market of this high fuel-efficiency technology in light-duty segments, several areas in which the chemical sciences could play a role are listed below.

While the chemical kinetics of the thermal autoignition process are relatively well understood, means of controlling the ignition timing in the engine cycle when operating in the HCCI mode are still elusive. Chemists and chemical engi-neers will need to help overcome this obstacle if HCCI is to be executable in automotive practice.

The most practical means of aftertreatment control of NO_x in the U.S. envi-ronment is the so-called NO_x trap, and these are highly susceptible to sulfur poi-soning. Even with the reduced, legislated levels of sulfur in Diesel fuel, the long-term effects of sulfur on NO_x trap performance is a major concern. Any means to reduce this sulfur sensitivity would contribute to the long-term success of Diesel technology.

The health effects of Diesel PM are contentious, and there are strongly held opinions on both sides of the issue. What is clear is that with the aforementioned advances in Diesel fuel-injection technology, PM mass has indeed decreased, but the PM number has become relatively greater. More decisive scientific conclu-sions on PM toxicity in general could contribute to moving this debate toward the establishment of realistic positions on the issue.

THE CO₂ TECHNOLOGY PLATFORM

Ruben G. Carbonell
North Carolina State University

The billions of pounds of organic and halogenated solvents used each year in chemical and materials manufacturing contribute significantly to the total discharge of volatile organic carbon (VOC) to the atmosphere and to the contamination of water and soil. In addition, drying operations utilize huge amounts of energy that is generated mostly by the burning of fossil fuels. As a result, there is great interest in finding novel alternative solvents that would result in cleaner, more efficient processes to enable a sustainable chemical and industrial manufacturing base.[1]

High-pressure carbon dioxide is one of the leading alternative solvents for many applications because of its unique physical and chemical properties.[2,3] It has a very accessible critical temperature and pressure (31°C and 73.8 bar) that enable its use in the liquid, supercritical, or gaseous state near room temperature. It is also highly compressible, so that its density and other physical properties can be varied over a wide range in order to control its solubility properties. Because it is generated as a by-product in the production of hydrogen, ammonia, and ethanol and because it is present in large quantities in many underground reservoirs throughout the world, it is relatively inexpensive compared to organic or halogenated solvents. So much carbon dioxide is currently generated by power plants and ends up in the environment, that even if all chemical and materials processes used CO_2 instead of organic or aqueous solvents there would still be no need to generate CO_2 from burning fossil fuels. However, the subsequent reduction in VOC and organic solvent emissions into the soil and aquifers would be quite significant.

In the supercritical state, CO_2 has a liquid-like density but a gas-like diffusivity, so mass-transfer and diffusion processes are greatly enhanced in this region. In its supercritical state it is also highly compressible, so that the density rises from approximately 0.5 g/mL at the critical temperature to about 0.9 g/mL at 10 °C. In that same range, the viscosity and surface tension remain at least an order of magnitude lower than those of water and most organic solvents. At these low temperatures the increased density enhances the solubility properties for organic compounds while the vapor pressure drops significantly from the critical

---

[1]Taylor, D. K.; Carbonell, R. G.; DeSimone, J.M. *Annual Reviews of Energy and the Environment* **2000,** *25,* 115-146.

[2]McHugh, M. A.; Krukonis, V. J. *Supercritical Fluid Extraction, Principles and Practice;* Butterworth-Heinemann, Oxford, UK; 1993; 2nd Edition.

[3]Hyatt, J. A., *J. Org. Chem.* **1984,** *49,* 5097-5101.

pressure, thus decreasing the potential cost of processes utilizing liquid instead of supercritical CO_2.

When considered as a solvent for chemical reactions, carbon dioxide offers a major advantage because it is essentially inert over wide ranges of temperatures and pressures. This prevents CO_2 from participating in any chemical reactions that can contaminate the product or terminate important elementary steps in a reaction that may control molecular structure or molecular weights. In addition, if used as a solvent or as a plasticizer, its high diffusivity and high volatility allow it to evaporate quickly and completely, eliminating all chances of contaminating the desired product. The high volatility also decreases the costs associated with solvent removal from a solid substrate when compared to water and organic solvents.

Many of these properties of CO_2 have been known for years,[2] but aside from some small specialty applications such as the extraction of caffeine from coffee beans and the fractionation of some polymeric compounds, CO_2-based processes have not made major inroads in industry. Over the last decade, interest in the use of CO_2 as a solvent has seen a great resurgence as a result of the discovery of some unique solubility properties associated with CO_2 that have enabled the synthesis of fluoropolymers in carbon dioxide as well as the rational design of surface-active materials that are soluble in CO_2.

Polymerization reactions are carried out industrially by a number of various methods, many of which involve organic solvents and water in the form of emulsions. These solvents often participate in the reaction, leading to unwanted side reactions. The polymer must then be dried, resulting in the expenditure of large amounts of energy, often leaving the product with significant amounts of unreacted monomer, and residual solvent. There are tremendous advantages in using CO_2 as a solvent in these types of reactions because the product ends up completely dry at the end of the process and any residual monomer could be extracted prior to product recovery. DeSimone and coworkers[4,5,6] showed that liquid and supercritical CO_2 could be used as a solvent for the solution polymerization of fluorooctyl acrylate and the precipitation polymerization of fluoroolefins such as tetrafluoroethylene with extremely high conversions, high molecular weights, narrow molecular weight distributions, and high purities. These discoveries led to the commercial development by DuPont of a novel, continuous polymerization process for Teflon based on this technology.[7] A pilot plant is currently operating in Bladen County, North Carolina, the first of what is likely to be a long-term investment by the company in CO_2-based polymerization processes.

[4]DeSimone, J. M.; Guan, Z.; Elsbernd, C. S. *Science* **1992**, *257*, 945-947.
[5]Romack, T. J.; Combes, J. R.; DeSimone, J. M. *Macromolecules* **1995**, *28*, 1724-1726.
[6]Romack, T. J.; DeSimone, J. M.; Treat, T. A. *Macromolecules* **1995**, *28*, 8429-8431.
[7]McCoy, M. *Chem. & Eng. News* **1999**, *77(4)*, 11-13.

Because of the high solubility of fluorocarbons (and siloxanes) in CO_2, it is not surprising that the first commercial direct synthesis of polymers in CO_2 was achieved with fluorinated materials. Siloxanes are also fairly soluble in carbon dioxide, but most other polymeric materials are not. CO_2 is an excellent solvent for high-vapor-pressure materials, but it is not a good solvent for high-molecular-weight, polar, or inorganic materials. To enhance the capability of CO_2 to dissolve these types of materials, it is necessary to use cosolvents or to design surfactants that are soluble in CO_2. The recognition that high-molecular weight fluoropolymers and siloxanes are soluble in CO_2 enabled the design of such surfactants via the covalent coupling of highly CO_2-philic moieties such as poly(fluorooctyl acrylates) with oleophilic or hydrophilic moieties. DeSimone and others[1,8,9,10] have demonstrated that these CO_2-soluble surfactants are able to form micellar aggregates in CO_2, and that these aggregates are able to enhance significantly the solubility of high molecular weight, oleophilic, and hydrophilic species. These CO_2-soluble surfactants are also able to stabilize latex particles in CO_2, and they can be used to form water-in-CO_2 and CO_2-in-water emulsions. It is interesting that the micellization process in compressed CO_2 is reversible.[11,12] At low CO_2 pressures, the surfactants dissolve in micellar form. However, as the pressure is increased, the micelle size (aggregation number) decreases until the surfactant solubility is so high that the micelles break up. Reduction of the pressure leads to reaggregation and recovery of the micellar structure. The solubility and functionality of these surfactants depend strongly on the ratio of CO_2-philic to CO_2-phobic groups, their chemical structure, and the morphology of the molecule (block or random copolymers). As a result, there are many ways of controlling system properties by careful control of the chemistry as well as the process conditions. One of the more natural applications for the use of these surfactants is in cleaning applications with CO_2 as the solvent. At least one commercial liquid CO_2-based dry-cleaning process[1] is currently in the marketplace, providing an outstanding alternative to perchloroethylene, in both cleaning quality and compatibility with garments, as well as in environmental and human health considerations.

The ability to synthesize CO_2-soluble bifunctional materials based on fluoropolymers or siloxanes has enabled many other potential applications for these novel material.[1] One of the more exciting opportunities is in the area of

[8]Maury, E. E.; Batten, H. J.; Killian, S. K.; Menceloglu, Y. Z.; Comes, J. R.; DeSimone, J. M. *Am. Chem. Soc. Div. Polym. Chem.* **1993**, *34(2)*, 664-665.

[9]DeSimone, J. M.; Maury, E. E.; Menceloglu, Y. Z.; McClain and, J. B.; Romack, T. J. *Science* **1994**, *265*, 356-359.

[10]McClain, J. B.; Betts, D. E.; Canelas, D. A.; Samulski, E. T.; DeSimone, J. M. *Science* **1996**, *274*, 2049-2052.

[11]Fulton, J. L; Pfund, D. M.; Capel, M.; McClain, J. B.; Romack, T. J.; Maury, E. E.; Combes, J. R.; Samulski, E. T.; DeSimone, J. M; Capel, M. *Langmuir* **1996**, *11*, 4241-4249.

[12]Buhler, E.; DeSimone, J. M.; Rubinstein, M. *Macromolecules* **1998**, *31*, 7347-7355.

coatings, with particular emphasis on microelectronics fabrication. This is an area in which the unique properties of CO_2 can lead to both process and environmental advantages.

For decades, photolithography has utilized a variety of organic and aqueous solvents to deposit, develop, and strip photoresists on silicon wafers. Currently, commercial photolithographic processes utilize chemically amplified photoresists and photoacid generators (PAGs) that are deposited from organic solvents using a spin coating technique. The patterns are then developed using an aqueous alkaline solution of tetramethylammonium hydroxide. Once the resist patterns have been etched into the underlying substrate, the photoresist is removed using a solvent- or water-based stripping step. Although these wet processes have served the industry well for years, as feature sizes continue to be reduced there are compelling reasons to reevaluate the use of conventional solvents and aqueous solutions and to implement a new "dry" technology.

To enable the manufacture of feature sizes with dimensions of 130 nm and less, the microelectronics industry uses photolithographic exposure tools operating at very short wavelengths of light (365 nm, 248 nm, 193 nm, and soon 157 nm). Lithography at 157 nm is particularly challenging since only a few classes of polymers have the requisite transparency at this wavelength. Thus most conventional polymeric materials are ill suited as 157-nm resists because the radiation cannot penetrate through films of the required thickness. Although most polymeric materials have a high optical absorbance at 157 nm, highly fluorinated polymers are relatively transparent, thus establishing that future generations of 157-nm resists will need to be highly fluorinated to be useful. Fortunately, these same glassy fluoropolymers are very soluble in liquid and supercritical CO_2.

We are attempting to design novel polymers and processes so that we can deposit the photoresist from liquid CO_2, develop the image after exposure with supercritical CO_2, and strip the post-etched material with supercritical CO_2 at higher pressures. Preliminary results have been obtained with a random copolymer of 1,1-dihydroperfluorooctyl methacrylate (FOMA) and 2-tetrahydropyranyl methacrylate (THPMA). With less than 30 mol % of the THPMA monomer, this photoresist in its fully protected form was found to be soluble in *compressed* liquid CO_2 at moderate conditions. In order to spin-coat films using liquid CO_2, the photoresist had to be soluble in liquid CO_2 at liquid-vapor equilibrium conditions where there is a meniscus that can allow the CO_2 liquid phase to spread and wet the wafer and to evaporate in a controlled manner leaving a uniform film of the photoresist on the wafer substrate. To achieve the desired solubility, all spin coating was performed under subambient temperatures (6-10 °C) that increase the density and solvating strength of CO_2 nearly 40% relative to room temperature. This was adequate to dissolve the photoresist at the desired liquid-vapor equilibrium conditions. Two fluorinated analogs of conventional PAGs were also designed to provide suitable solubility in liquid CO_2.

Polymer solutions for spin casting were prepared by dissolving 20 wt % of

PFOMA-*r*-THPMA in liquid CO_2. This liquid CO_2-photoresist-PAG solution was maintained in a high-pressure view cell equipped with sapphire windows so that complete dissolution could be confirmed visually. The spin coating was carried out in a specially designed and built high-pressure spin coating tool.[13,14] The pressure in the spinning chamber was maintained at precise values ranging from 5 to 15 psi below the equilibrium vapor pressure of the CO_2-photoresist solution. Control of this chamber pressure was a key parameter in the determination of film quality and thickness, as the pressure differential was what governed the evaporation rate of the liquid CO_2. Precision control of the evaporation rate, spinning speed, and system temperature were required to ensure consistent, uniform coatings. Extensive work was conducted to determine the ideal range of rotational speeds, pressures, and concentrations to be used to produce uniform, lithographic quality films, resulting in films with 3% variation in film thickness over the entire wafer and a root mean square (rms) roughness of 0.4-0.5 nm. The photoresist and PAG formulations were exposed and imaged at 248 and 193 nm, resulting in highly encouraging feature sizes of 0.8 µm. Even though this technology is not yet ready for full commercial implementation, it does point the way toward the development of a new lithographic process that is totally "dry" and compatible with existing cluster tools in the industry to couple to vacuum operations. All of this is accomplished with a single environmentally friendly solvent.

In other related areas of microelectronics fabrication we have also succeeded in developing a high-pressure free meniscus coating apparatus capable of depositing monolayer to 300 Å thin films of perfluoropolyethers with great uniformity and roughness characteristics.[14] We believe that the low viscosity of CO_2, coupled with its excellent wetting properties, will enable whole new classes of thin-film coating operations that will at the same time be environmentally responsible. These are likely to be important, not just for microelectronics applications but also for biomedical and nanotechnology formulations. Even though there are still many technical and economic barriers to the total acceptance of these technologies, we believe that environmental pressures as well as technical requirements for pure component systems with high uniformity will over time help "dry" CO_2-based processes play an increasingly important role in industrial environments.

---

[13]DeSimone, J. M.; Carbonell, R. G. U.S. Patent 6,001,418; 2001.

[14]Hoggan, E. N.; Novick, B. J.; Carbonell, R. G.; DeSimone, J. M. *Semicon. Fabtech* **2002**, *16*, 169.

SMART, SUSTAINABLE GROWTH

Uma Chowdhry
The DuPont Company

This paper describes DuPont's approach to addressing the dual challenge of achieving business growth through new products that create a high standard of living while also protecting the environment for future generations. We have made environmentally sustainable growth (Box 1) through technical innovation our primary challenge, and this ideology serves as a great motivator for our people and our businesses.

DuPont's bold and uncompromising commitment to sustainable growth with

---

BOX 1
Smart, Sustainable Growth

DuPont's commitment:

Creating shareholder and societal value while decreasing our environmental footprint along our value chains.

"Environmental footprint" = injuries, illnesses, incidents, waste & emissions, and depletable forms of raw materials and energy

---

the intent of integrating economic, environmental, and social factors forms the foundation for our future. This simple statement represents an enormous challenge. It requires that we focus our efforts and also track our environmental impact. At DuPont we are measuring the way we "create shareholder value while decreasing our environmental footprint along our value chains." The term "footprint" includes raw materials, energy, emissions and waste, as well as injuries, illnesses, and environmental incidents. We are integrating this concept into all business decisions, into all local actions at the community level, and into helping lawmakers make fact-based decisions in enacting new pieces of legislation. This is a tall order, but if we are to protect the environment for future generations, we must take on this challenge.

Our core values (Box 2), which have stood the test of time, provide us with the organizational culture to take on this enormous challenge of sustainable growth. With strong leadership, we believe we can stay the course. Since the days of our company's founders, we have had a mindset of "zero injuries" to our people. We are extending that same mindset to environmental excellence and

---

BOX 2
DuPont's Core Values

- Safety, health: goal is "0"
- Environmental excellence
- Business ethics: highest standards
- People: treat with dignity and respect

Core values define who we are and are "nonnegotiable."

---

have set demanding stretch goals for ourselves to remain at the forefront, leading the challenge for the chemical industry, and influencing our customers and competitors to adopt the same goals of zero injuries to people or to the environment.

As DuPont enters its third century and we reflect on the chemical industry over the last 200 years (Box 3), it is clear that we have contributed very significantly to improved standards of living and to material prosperity for people around the world through better housing, apparel, transportation, and food. The industry has created both economic and societal value over the last two centuries. The commitment to continue this pursuit now has to be accompanied by a commitment to protect our environment.

The chemical industry's challenge lies in cleaning up contaminated water, reducing toxic air emissions, and reducing energy consumption and waste that leads to soil contamination (Box 4). Today the most pressing issues for water contamination are heavy metals, nondegradable biologically active substances, and persistent bioaccumulative toxins. Emissions from the chemical industry have contributed to ozone-depleting materials and greenhouse gases, causing global climate change. Our land is contaminated in certain areas with toxic compounds,

---

BOX 3

Over the last 200 years:

Chemical Industry has contributed to material prosperity in housing, apparel, transportation, and food, by utilizing our knowledge of chemistry to provide

Better Things for Better Living

But it came at a price.

BOX 4
Chemical Industry

- Water Contamination
  - ○ Heavy metals
  - ○ Biologically active, nondegradable substances
  - ○ Persistent bioaccumulative toxins
- Air Emissions
  - ○ Greenhouse gases
  - ○ Ozone-depleting materials
- Soil Contamination
  - ○ Toxic metal compounds
  - ○ Heavy metals
  - ○ Biological waste

---

heavy metals, and biological waste. All of these harmful consequences of the material prosperity we have enjoyed are compromising the health of present and future generations.

Looking broadly at the spectrum of industries that generate 95% of water, air, or soil emissions, we see that they range from chemicals, metals and mining, food, and paper to petroleum, utilities, and equipment manufacturers. The chemical industry ranks first, second, and fifth in the generation of water contaminants (Figure 1), air toxics (Figure 2), and soil contaminants (Figure 3). It is therefore incumbent upon the chemical industry to take a strong stand on reducing the environmental footprint that it has generated and could potentially create in the future.

We must progress from an attitude of just complying with federal regulations to one where we earn the public's trust and move on to sustainable development. This corresponds to a journey: from compliance, to earning the public's trust, to sustainable development.

A quick review of the results over the past decade at DuPont demonstrates our commitment to sustainability (Box 5). While production volume grew 35%, we kept energy consumption flat through innovation in our manufacturing processes. Impressive reductions were made in air carcinogens (86%) and toxics (73%), in greenhouse gas emissions (63%), and in hazardous waste generated (40%) as well as reduction in deep well disposal (82%). Conservatively these environmental improvements have also led to over $1 billion in savings for DuPont. What is good for the environment can also be good for business.

DuPont is transforming itself as it enters its third century. Using a three-pronged strategy of integrated science, knowledge intensity and "Six Sigma" productivity (see below), we are focused on sustainable growth (Box 6).

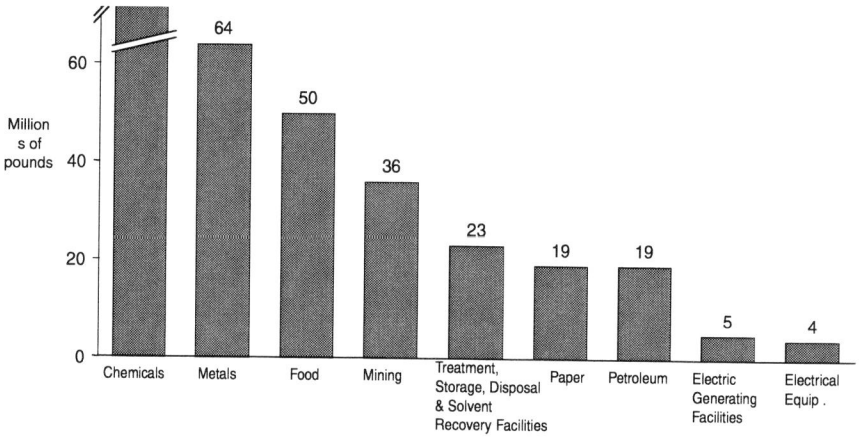

FIGURE 1 Chemical water emissions by industry. These industries contribute 95% of the total emissions (1999 data). Source: DuPont Consulting Solutions analysis of EPA Toxics Release Inventory, 1999.

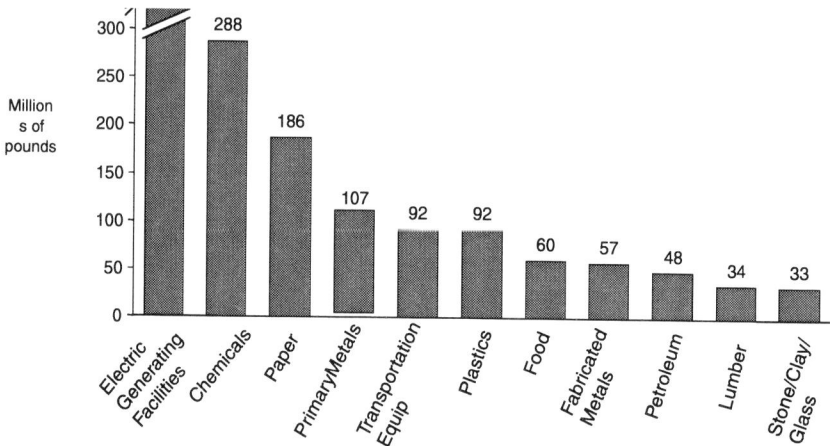

FIGURE 2 Chemical air emissions by industry. These industries contribute 95% of the total emissions (1999 data). Source: DuPont Consulting Solutions analysis of EPA Toxics Release Inventory, 1999.

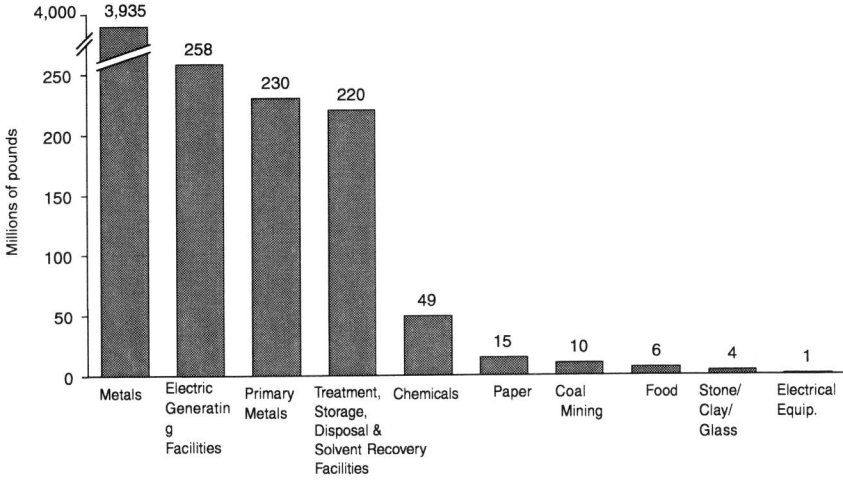

FIGURE 3 Chemical soil contamination by industry. These industries contribute 99.6% of the total emissions (1999 data). Source: DuPont Consulting Solutions analysis of EPA Toxics Release Inventory, 1999.

BOX 5
DuPont's Environmental Footprint Reduction

Performance over last 10 years

Safety & Health	World Leader
Air Carcinogens	↓ 86%
Air Toxics	↓ 73%
Hazardous Waste	↓ 37%
CO_2 equiv Greenhouse Gases	↓ 63%
Total Emissions	↓ 76%
Total Waste Generated	↓ 38%
Production Increase	↑ 35%
Hazardous Waste	↓ 40%
Deep Well Disposal	↓ 82%

Volume increased 35%, while energy use was flat!!

BOX 6
DuPont's Strategy for Smart, Sustainable Growth

- Integration of science and technology platforms for new products and processes

   ⇒ less environmental impact
- Knowledge intensive services combined with products

   ⇒ increased shareholder value added per pound (SVA/lb)
- Focus on asset productivity

   ⇒ less waste

DuPont's heritage is its ability to create value with science and technology. In the twentieth century, we combined chemistry and engineering to create new materials such as nylon, Teflon, Lycra, and Kevlar that are household words today and have brought the world fashion, comfort, and protection. We are transforming ourselves by building on our heritage of innovation to begin integrating biology and chemistry into creating value. We are investing in building our capability in biotechnology because we see enormous potential in using this platform of new technologies to ensure sustainable development.

Another pathway to sustainability is to generate more business value but with fewer pounds of material. We have a metric called shareholder value added per pound of product produced (SVA/lb). SVA is the value created above the cost of capital. By selling high-value services coupled with our products, we can enhance progress toward high SVA/lb. When coupled with other financial metrics, SVA/lb provides an indicator of future sustainability for different growth strategies.

The focus on asset productivity at DuPont is relentless. Improving yields, uptime, and throughput helps us delay capital expenses and reduce raw material usage and waste. We have adopted the Six Sigma methodology that uses statistically significant, data-based understanding to reduce defects in our manufacturing processes. This methodology, coupled with process innovation, has led to lower raw material usage, lower energy consumption, lower emissions, and lower waste with annual pretax savings of over $1 billion over the past three years. Thus, through technological innovation, we have made significant progress toward creating economic and societal value while reducing our environmental footprint (Box 7).

Several examples will illustrate the use of technical innovation to provide "greener" products and processes. These are elimination of CCl_4 waste during phosgene production; production of Tyvek envelopes from waste; Smart automo-

---

BOX 7
Towards Greener Products & Processes

- Eliminate waste at source vs end of pipe treatment
- Recycle waste to useful products
- Reduce emissions in our plants and for our customers
- Develop alternate forms of energy
- Use renewable resources for chemical production
- Increase use of biobased solutions
- Increase bioefficiency

---

tive finishes; biobased routes to chemicals, polymers, and fibers; and a biomass refinery.

The first example deals with reduction of carbon tetrachloride formation in the production of a hazardous chemical, phosgene (Figure 4). Phosgene serves as an intermediate for some of our specialty fibers and agricultural intermediates. The State of New Jersey required us to reduce carcinogenic CCl_4, which forms as a by-product when chlorine and carbon monoxide are heated at high pressure. Our goal was to reduce CCl_4 formation from 500 to 100 ppm. Our approach was

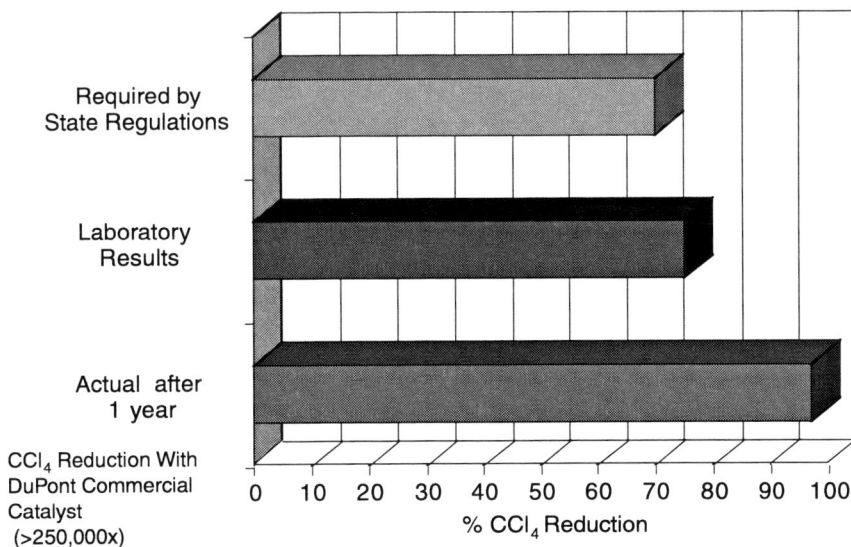

FIGURE 4 Reduction of CCl_4 through improved catalyst.

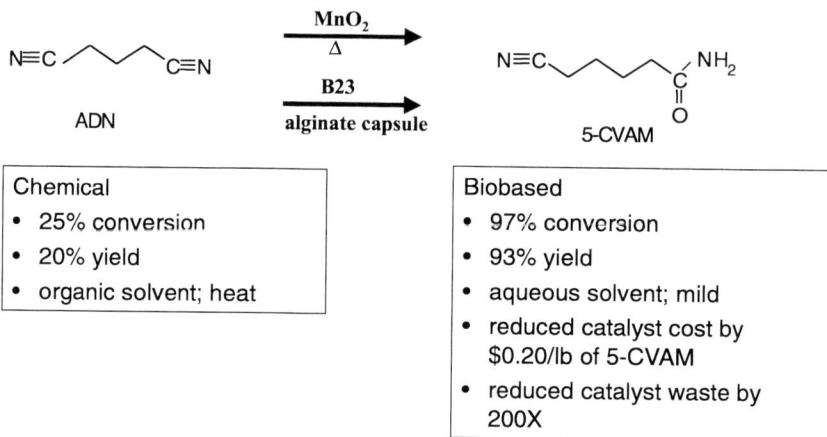

FIGURE 5 Chemical vs. bio-based route to an agricultural intermediate.

to find superior catalysts that, in fact, reduced CCl_4 formation from 500 to less than 10 ppm, exceeding regulatory requirements. We are licensing this technology to other phosgene manufacturers to ensure that this development is available not only for DuPont but for other manufacturers as well.

Another example shows the ingenuity of our technical community in being able to use recycled water and milk jugs made of high-density polyethylene into Tyvek envelopes that are lighter than conventional Tyvek used for housewrap during construction (Figure 5). The added benefit to the airlines of lower-weight cargo is the lower use of fuel and lower emission of greenhouse gases. Polyethylene-based Tyvek is a non-woven material used to wrap houses under construction to improve insulation. This use of Tyvek reduces energy consumption by tenfold each year for the homeowner.

In our automotive coatings business, we lowered waste by developing a coating with very high-solids content. The advanced analytical tools available today allowed us to measure and control the distribution of low molecular weight material in the coating. This, combined with hybrid cross-linking, provided a thinner coating with superior durability and scratch resistance coupled with smoothness and higher gloss. Not only did we reduce waste at our manufacturing site by decreasing the use of solvents, but we were able to help our customer Daimler-Chrysler reduce its volatile organic compounds by 1.5 pounds per vehicle, hazardous air pollutants by 0.45 lbs per vehicle, odor emissions by 86%, and total raw material usage by 20%. Sustainable development should allow all parties along a producer's value chain to benefit from product and process improvements.

Switching to examples of bio-based solutions (Figure 5), we demonstrated the use of an enzyme—a renewable resource—in the production of 5-

cyanovaleramide, an agricultural intermediate important for citrus growers and some vegetable farmers. The results using a bio-based catalyst vs. a conventional oxide catalyst show dramatic improvements. The chemical process operated at 25% conversion and 20% yield and used an organic solvent in a typical chemical reaction. The use of a specific enzyme allows 97% conversion and 98% yield and employs an aqueous solvent under mild conditions. The bio-based operation also allows reduction of catalyst cost by $0.25 per pound and results in twentyfold waste reduction. This impressive example confirms that it is possible to use renewable resources to create economic value while creating societal value and reducing our dependence on petroleum-based feedstocks.

We embarked on a large program five years ago to demonstrate that we can use our technical innovation power, coupled with a partner's capability in engineering enzymes, to make bulk chemicals and fibers cost effectively. Our goal was to produce a specialty fiber we call "Sorona" (Figure 6), which incorporates the attractive properties of nylon, Dacron and Lycra, resulting in superior softness, vibrant color, UV and chlorine resistance, and stain resistance coupled with stretch and recovery. Market test development shows that consumers find this combination of functionalities very attractive.

Chemical routes to polypropylene terephthalate-based fibers, which we have branded Sorona, use hazardous chemicals such as ethylene oxide and carbon monoxide and are subject to the environmental problems of a typical chemical process (Figure 7). We undertook the enormous challenge of producing 1,3-propanediol (3G) from glucose in one step as shown in Figure 8.

First attempts resulted in very low yield. Working with Genencor, our team of scientists and engineers has genetically modified an enzyme to gain a 120-fold improvement in yield over five years (Figure 9). This "tour de force" of modifying an *E. coli* host to achieve high conversion and selectivity has demonstrated to us and to the world what biocatalysis can do for the chemical industry. We have scaled up the process and are currently operating a pilot plant.

The impressive results we have demonstrated with bio-3G prompted the Department of Energy (DOE) to award a multimillion dollar grant over four years to DuPont, Diversa, and the National Renewable Energy Laboratory (NREL) to use the biomass from our 3G process as fuel. Our target is to demonstrate the use of renewable resources (corn sugar) and simultaneous production of fuel from the biomass—a "biorefinery" as an economically viable concept (Box 8).

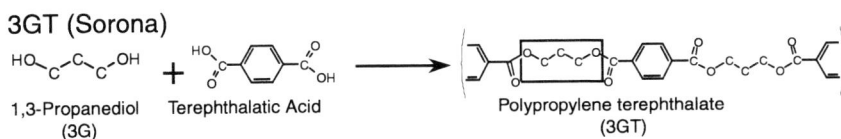

3GT (Sorona)

1,3-Propanediol   Terephthalic Acid   Polypropylene terephthalate
(3G)   (3GT)

FIGURE 6 Sorona (polypropylene terephthalate), an advanced polymer-fiber.

Shell:

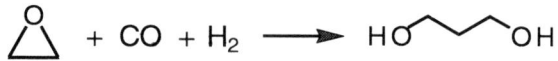

Degussa:

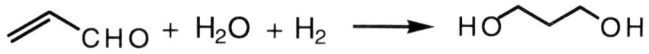

FIGURE 7 Traditional chemical routes to 3G.

In Nature:

Two microorganisms convert sugar to 3G stepwise.

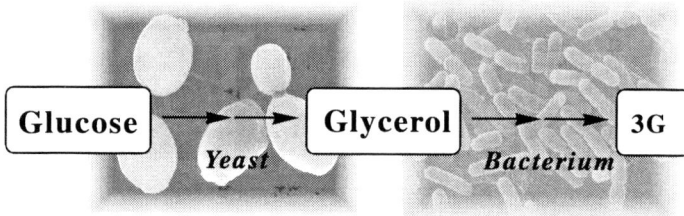

For an Industrial Process: A single microorganism is desired

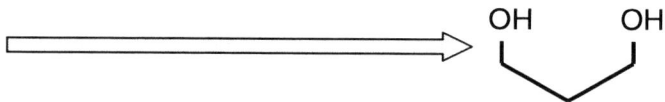

FIGURE 8 Nature combined with metabolic engineering to produce 3G.

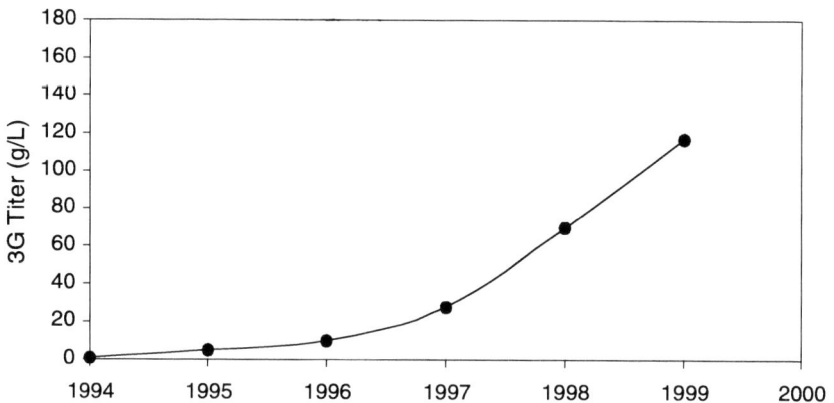

FIGURE 9 Bio-3G titer history.

BOX 8
Integrated Corn-Based Biorefinery (ICBR)

Corn + stover → fermentable sugars → PDO (3G) + Ethanol

Measures of success:
- Petroleum usage down 90%
- (equiv $1MM barrels of oil/yr per ICBR)
- Climate change gases down 90%
- Reduced investment by $120MM per plant
- Returns on investment > 10% after tax income

DuPont's commitment to sustainable development begins with a strong leadership commitment to stay the course through economic cycles, to set stretch goals for each decade and to drive innovation. Specific 2010 goals for energy are to

- derive 25% (up from 10% today) of corporate revenue from renewable resources;
- source 10% of our energy needs from renewable resources;
- reduce greenhouse gases by 65% vs. 1990; and
- keep total energy use flat vs. 1990.

Our commitment to use our powerful technology base coupled with complementary skills from universities, government labs, and/or other companies to ensure sustainable development is steadfast. Further examples of technical programs to illustrate this commitment include fuel cells, sensors, and catalysts for controlling auto emissions, waterborne coatings, supercritical solvents, and recyclable polymers.

Major environmental trends that we see for land, air, water, and transportation of environmentally hazardous materials are shown in Box 9. These trends require that we get ahead of these issues and lead the chemical industry in the reduction of toxic metal (e.g., Sb, Sn, As) compounds, greenhouse gases, mercury emissions, and sulfur from gasoline and diesel, and find ways to control and sequester CO_2. Reduction of arsenic, as well as nitrates and ammonia, in drinking water is necessary. It is also imperative in these days of terrorism that we reduce transportation and storage of hazardous materials and continue our drive to develop inherently safer processes.

DuPont has prospered in the last two centuries. As we enter our third century (Figure 10), we are committed to use the power of biology coupled with our

BOX 9
Major Environmental Trends

Land
  • Growing regulations for toxic metal compounds (Sb, Sn, As, etc.)
  • Increasing volume of biological waste
Air
  • NO_x, SO_x reduction (75% below 1997 level by 2007)
  • Hg emissions reduction (90% from 1997 level by 2007)
  • Sulfur in gasoline reduction (30 ppm by 2005)
  • Sulfur in diesel reduction (15 ppm by 2007)
  • CO_2 control and sequestration
  • Phase-out of brominated organics
Water
  • Phase-out of methyl *tertiary*-butyl ether (MTBE)
  • Arsenic reduction in drinking water
  • NH_3, NO_3 limits in waste water
Security
  • Movement away from storage of hazardous substances
  • Embed security code into responsible care
  • Introduce inherently safer processes

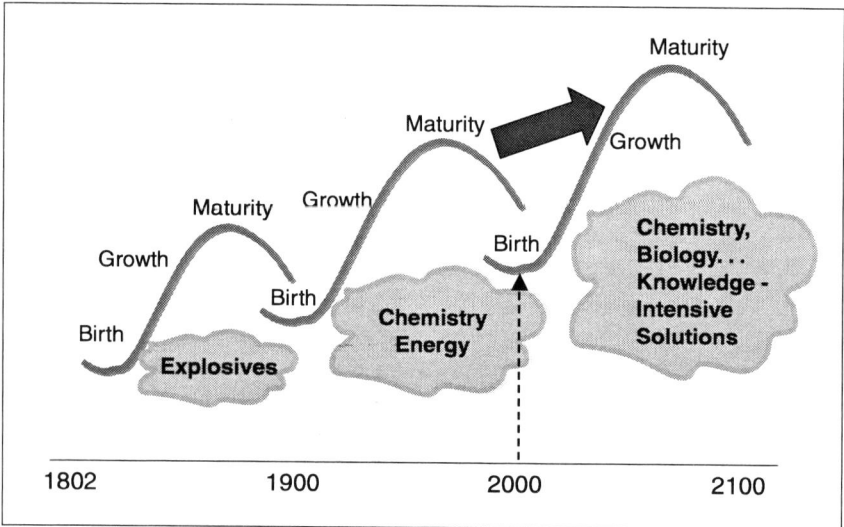

FIGURE 10 The transition to smart, sustainable growth.

traditional strengths in chemistry and engineering to what we hope will become the "century of sustainability." The ultimate goal of sustainability has to be conservation of global ecosystems for the future of our planet and of humanity. We cannot settle for anything less.

THE ORIGIN AND NATURE OF TOXIC COMBUSTION BY-PRODUCTS

Barry Dellinger
Louisiana State University

Introduction

Although combustion and thermal processes are necessary to provide for the essential needs of our existence, they are intrinsically "dirty" and emit a variety of air pollutants. Some of these pollutants are well known, well understood, and subject to significant control.

However, combustion is a complex process that results in formation of many pollutants that are not well characterized as to their nature or origin. As a responsible society, it is incumbent upon us to examine these issues, determine their importance, and endeavor to eventually resolve and address each of them.

Combustion-related air pollution can be classified in many ways. Table 1 presents one such classification that may assist in identifying and prioritizing research needs.

We actually have a pretty good idea of the identity of most of the emissions from combustion on a mass basis; on the order of 99.9% of emissions from a reasonably controlled combustion source are carbon dioxide, carbon monoxide, and simple hydrocarbons such as methane and ethane. However, the remaining fraction is a complex myriad of pollutants that is not fully characterized and can contain toxic species.

TABLE 1 Categorization of Combustion-Generated Air Pollutants.

Category	Examples
Smog precursors	CO
Acute toxics	NO_x, volatile organic compounds (VOCs)
Toxic air pollutants	Butadiene, Polycyclic aromatic hydrocarbons (PAHs)
Endocrine disrupting chemicals	Dioxins, oxy-PAHs
Halocarbons	Chlorinated hydrocarbons (CHCs), brominated hydrocarbons (BHCs)
Fine particles	Metals and organic constituents
Persistent radicals	Semiquinones

TABLE 2 Chemical Reactions Zones in Combustion Systems.

Zone	Reaction Conditions	Decomposition Mechanisms	Formation Mechanisms
1 Pre-flame	T = 200-1000°C $t_r \ll 1$ s $[O_2]\sim 50\%$ EA	Molecular eliminations, bond fission, bimolecular radical attack	Molecular eliminations, complex radical-molecule pathways, recombination-association reactions
2 Flame	T = 1000-1800°C $t_r \sim 0.01$s $[O_2] \sim 50\%$ EA	Bimolecular radical attack, bond fission, molecular eliminations	Complex radical-molecule pathways, molecular eliminations, recombination-association reactions
3 High-temperature thermal	T = 600-1100°C $t_r = 1$-10s $[O_2] = 50$-100% EA	Molecular eliminations, bond fission, bimolecular radical attack	Recombination-association reactions, complex radical-molecule pathways, molecular elimination
4 Gas quench	T = 80-600°C $t_r \sim 10$s $[O_2] = 3$-9% EA	Molecular eliminations, bond fission	Recombination-association reactions
5 Surface catalysis	T = 1000-1800°C $t_r = 10$s to 10 min $[O_2] = 3$-9% EA	Surface-catalyzed decomposition	Surface-catalyzed synthesis

An analytical chemist would be interested primarily in characterizing these emissions by developing new analytical techniques and continuous monitoring apparatus. A combustion scientist might be most interested in identifying the origin and mechanism of their formation. A combustion engineer might focus on developing methods for their mitigation through design of control technology or combustion modification to prevent their formation.

Table 2 presents a combustion system from the viewpoint of a combustion scientist. It identifies reaction zones, the conditions that exist within these zones, and classifications of reactions that can occur under these conditions.

The vast majority of these pollutant-forming pathways involve free radicals. It is generally assumed that these radicals are formed in the high-temperature flame zone of combustion systems. However, reactions occurring in the post-flame, thermal zone (Zone 3) and the gas-quench and surface-catalysis zones (Zones 4 and 5), may also form radicals responsible for pollutant formation. In some cases, the radicals may be stable and act as pollutants themselves.

Small reactive radicals (e.g., HO·, H·, O·) have lifetimes of less than a microsecond. Organic radicals are less reactive and may have lifetimes of several microseconds. Resonance-stabilized organic radicals, such as cyclopentadienyl and propargyl, can be even more stable and less reactive, with lifetimes in the millisecond range. Catalytic cycles similar to those well studied by the tropospheric chemistry community can result in measurable steady-state concentrations of radicals that exist for several seconds after the combustion event that initiated their formation. Finally, recent evidence suggests that semiquinone-type radicals contained in some types of particles may persist indefinitely.

Combustion Chemistry Research Opportunities

While some sources and mechanisms of combustion-generated air pollution have been the subject of considerable study, other sources are poorly characterized and not very well understood. Examples are:

1. flares and plumes,
2. soot and PAH formation by resonance-stabilized radicals,
3. endocrine disrupting chemicals (EDCs) and oxy-PAH,
4. gas phase reactions of halogenated hydrocarbons,
5. surface-mediated pollutant formation, and
6. particulate-stabilized free radicals.

Flares and Plumes

Industrial flares and plumes represent a potentially significant source of air pollution that are poorly characterized and controlled. Conditions are ideal for post-flame thermal reactions and photolytic reactions at elevated temperatures (photothermal reactions). The elevated temperatures in flares and plumes (~50 to 600 °C outside of the visible flame) can result in accelerated rates of formation of oxy-PAH and nitro-PAH. At higher temperatures within the flame zone of combustors, oxy-PAH and nitro-PAH are likely to be destroyed, but under the relatively mild conditions of flares and plumes, the rates of formation can be accelerated without their subsequent destruction. The elevated temperatures and exposure to solar radiation can result in fast photothermal reactions that lead to the formation of both combustion-type pollutants and photochemical pollutants. Research has shown that at elevated temperatures, the rate of absorption of solar radiation and photochemical quantum yield can increase up to tenfold.

Soot and PAH Formation by Resonance Stabilized Radicals

For many years, the reactions of small organic radicals, containing even numbers of carbons, such as vinyl, ethynyl, and butadienyl, have dominated the theory of molecular growth to form soot and PAH. However, it has been recognized recently that odd-carbon radicals such as propargyl and cyclopentadienyl are stabilized and play a significant, possibly dominant role. The properties and elementary reactions of these resonance-stabilized species are poorly characterized from the chemical viewpoint. They can undergo a variety of isomerization, recombination, and addition reactions for which rates have not been determined, and their impacts on PAH formation pathways have not been assessed.

EDCs and Oxy-PAH

It is now known that endocrine disrupting chemicals are emitted from combustion sources. Interest has focused on the emissions of polychlorinated dibenzo-*p*-dioxins and polychlorinated dibenzofurans (PCDD/F), which are also known carcinogens. However, oxy-PAH, epoxides, and other oxygenated species are known EDCs. These can also be emitted from combustion sources, although they are not well characterized. They are semipolar compounds that are difficult to analyze. Thus, improved methods of analysis are needed in conjunction with biological testing to determine the nature and quantity of EDC emissions from combustion sources.

Gas-Phase Reactions of Halogenated Hydrocarbons

Chlorinated and brominated materials are burned or thermally treated in a variety of combustion sources including hazardous and municipal waste incinerators, industrial processes, backyard trash burning, and accidental fires. Chlorinated materials are used in a wide range of applications and brominated compounds are fire retardants used in many devices including electronic circuits. Although there has been some research on the reactions of CHCs and BHCs in the past 20 years, too little is known about their reactions considering the magnitude of the environmental impact. Elementary reaction studies of gas-phase reactions of C_1 and C_2, CHCs, and BHCs are needed to understand their most fundamental reaction properties. Reactions of the chlorinated and brominated benzenes and phenols are important intermediate steps in the formation of PCDD/F. Recent kinetic models indicate that the gas-phase reactions may be quite important and elementary gas-phase reaction studies have been overlooked by researchers.

Surface-Mediated Pollutant Formation

Research has shown that the presence of catalytic surfaces and particles increases the yields and rates of formation of PCDD/F in combustion systems over the reaction temperature range of 200-600 °C. Transition metals such as copper can increase the rate of chlorination, molecular growth, and aromatic condensation reactions to form PCDD/F. Also, reactive species can attack a carbon matrix to chlorinate and fragment the carbon lattice-forming PCDD/F as well as other chlorinated hydrocarbons. Although research to date has focused on surface-mediated PCDD/F and chlorinated hydrocarbon formation, this same research suggests that many types of pollutants can be formed by similar processes. Postcombustion cool-zone formation of pollutants may explain an important combustion dilemma, which is how seemingly thermally fragile compounds can be emitted from combustion systems.

Particle-Stabilized Free Radicals

Recent research has shown that combustion sources can generate radicals that are stabilized by associated with particulate matter. This same particulate matter becomes a component of airborne PM2.5. (fine particulate matter (smaller than 2.5 microns in diameter). PM2.5 is known to initiate lung cancer and cardiopulmonary disease; however, the mechanism has not been identified. DNA and cellular assay results indicate that combustion and PM2.5 can cause radical induced damage to DNA. Based on electron paramagnetic resonance (EPR) studies, the responsible species appear to be semiquinone-type radicals. These studies reveal that radicals, heretofore thought to be too unstable to survive in the atmosphere, can be stabilized by association with particles and initiate biological damage.

Conclusions and Recommendations

Research on the environmental aspects of combustion is inherently multidisciplinary. Fundamental research is needed within specific areas; instrument development is needed to facilitate this research; and interdisciplinary collaborations are needed to evaluate the health impacts of combustion-generated pollution.

Research Recommendations

• Research on photothermal and thermal reactions in flares and plumes including photothermal chemistry and spectroscopy as well as destruction and formation of toxic air pollutants

• Elementary reaction kinetic studies of resonance stabilized radicals and how they impact formation of PAH and PAH radicals

• Mechanistic studies of the partial oxidation of PAHs by thermal, photolytic, and photothermal pathways

• Elementary reaction kinetic studies of CHCs and BHCs with specific emphasis on the reactions of chlorinated phenol and other dioxin precursors; chemically activated displacement reactions; Cl•, Br•, H•, O•, and HO• reactions; and ab initio molecular orbital calculations

• Research on surface-catalyzed pollutant formation including transition

metal catalyzed dioxin formation, surface-catalyzed formation of CHC and toxic air pollutants, and catalytic destruction
• Efforts to characterize fine particles including speciation of toxic metals and organometallic surface binding
• Research on persistent radicals in the environment including characterization of their structure, mechanisms of formation, mechanisms of stabilization, and pathways of biological redox cycling

Instrument Development Needs

• Methods for study of fast surface reactions
• Surface analysis techniques for organometallic binding and radical characterization
• Dependable methods for studying elementary gas-phase reactions of organic radicals at elevated temperature
• Methods for studying the spectroscopic properties of high-temperature systems
• Techniques for metal speciation

Recommendations for Support of Research

• Development of microarrays for rapid screening of biological end points of complex mixtures
• Risk assessment methods for multiple pollutants from multiple sources
• Biochemical reactions of environmentally persistent free radicals

COMPUTATION AND ENVIRONMENTAL SCIENCE

David A. Dixon
Pacific Northwest National Laboratory

Environmental chemical science deals with issues of scale as much as any area of chemistry and the issues of scaling in space and time dominate environmental science. The goal of environmental science is to understand the current state of the environment based on our knowledge of the past and to use this information to be able to predict the future state. For example, given current practices for manufacturing, what will be their long-term environmental impact? Given potential environmental remediation strategies, what will these lead to? One does not want to use a remediation strategy that will have unforeseen consequences and introduce new environmental issues. No one wants to repeat the mistakes of the past, for example, the wide release of chlorofluorocarbons (CFCs) into the atmosphere that led to stratospheric ozone depletion. Although, we are interested in the results at large spatial and temporal scales, detailed insight into behavior at the molecular scale is key to understanding (1) how humans have impacted the environment, (2) how to remediate anthropogenic impacts on the environment, and (3) how to minimize future anthropogenic impacts on the environment.

Computing has revolutionized the way that we live and the way that we practice science. The entire research enterprise has been undergoing a revolution over the past two decades as it exploits the advances that are occurring in computer hardware and software and in new mathematical and theoretical approaches. This revolution is based on the utilization of high-performance computers (now massively parallel) to solve the complex equations that describe natural phenomena (e.g., the Schrödinger equation for electronic motion in molecules; Newton's equations of motion for the classical motion of hundreds of thousands of particles such as those in a protein). Modeling and simulation is now considered to be the third branch of science, bridging experiment and analytical theory. The role of simulation in the modern scientific and technical endeavor cannot be underestimated, and the use of effective modeling and simulation plays a critical role in modern scientific advances. There are a number of important roles that modeling and simulation play in the scientific enterprise. First, modeling, theory and simulation can enhance our understanding of known systems. Second, they can provide qualitative and quantitative insights into experimental work and guide the choice of which experimental system to study or enable the design of new systems. This is most useful if the simulation has been benchmarked on well-established systems to validate the approach. Third and finally, simulations can provide quantitative results to replace experiments that are too difficult, dangerous, or expensive and can extend limited experimental data into new domains of parameter space. For example, accurate thermochemical and kinetic calculations for the design of nuclear waste processing facilities and green chemical processes or for predicting tropospheric oxidation processes relevant to aerosol formation are needed due to missing experimental data. In addition, simulation allows one to explore temporal and/or spatial domains that are not accessible by present experimental methods. For example, it is now possible to explore different chemical reaction pathways not directly accessible by experiment to learn why they are not favorable or to find missing steps in a mechanism.

High accuracy from a simulation is important. A factor of 2 to 4 in catalyst efficiency may determine whether a chemical process is economically feasible or not, and a factor of 4 in a rate constant at room temperature (25 °C) corresponds to a change in the activation energy on the order of just less than 1 kcal/mol. For a 50:50 starting mixture of two components, a change in the free energy, ΔG, of less than 1.5 kcal/mol leads to a change in the equilibrium constant by a factor of 10, leading to a 90:10 mixture at 25 °C. The requirement for such accuracy means that we must be able to predict thermodynamic quantities such as bond dissociation energies (D_e or D_0^0) and heats of formation (ΔHf) to better than 1 kcal/mol and activation energies to within a few tenths of a kilocalorie per mole—a daunting computational task. Rapid advances in hardware, algorithm development, theory, and software are enabling computational scientists to attack larger and more complex problems with higher-accuracy and higher-fidelity models. Based on advances in computational science over the past two decades, we often know how to dramatically improve the quality of the simulation given sufficient computing resources. If we are to gain the maximum impact from simulations, one must aim for the highest possible accuracy in the simulations given the available resources, and one must continue to develop methods that can take advantage of the significantly increased computational resources to be available in the future. This latter point is critical due to the rapid evolution of computer hardware, driven mostly by the consumer industry.

It has recently been shown that computational chemistry methods can provide the accuracy required to reliably solve complex environmental problems but accuracy significantly increases the computational demands. Examples of how computational chemistry is being used to impact environmental science include the following:

- Accurate properties prediction for radionuclides, including actinides and lanthanides, to understand their migration in the vadose zone (e.g., the Hanford site), and their chemical behavior in waste tanks (e.g., Hanford and Savannah River)—such chemical reactivity information is needed for detailed subsurface and groundwater reactive transport models.
- Reliable prediction of thermodynamic and kinetic properties for chemical processes (e.g., reactions of chlorinated hydrocarbons on surfaces and in aqueous systems, atmospheric oxidation of organic precursors to ozone and aerosols) as well as for designing green chemical manufacturing processes.
- Molecular-level studies of chemistry in solution and at interfaces, including mineral interfaces (e.g., the behavior of metal ions in aqueous solution and on metal oxide or clay surfaces for vadose zone, tank, and groundwater remediation and catalysis)—a detailed understanding of redox (electron transfer chemistry) is broadly needed; studies of the interactions of biological molecules with surfaces for bioremediation arc also needed and being pursued.
- Reliable prediction of spectroscopic properties to aid in the interpretation of experiments for determining speciation in the environment (e.g., surface, subsurface, groundwater) and in tanks, as well as for chemical process control.
- Reliable prediction of chemical processes for carbon management including aerosol formation (organic oxidations, inorganic NH_3-H_2SO_4-H_2O chemistry, and nucleation processes) and sequestration and capture of CO_2 (e.g., in geologic formations and in the ocean).
- Fundamental molecular processes relevant to cell signaling pathways for biological remediation and risk assessment including protein-protein interactions and enzymatic reactions.

- Structural biology and functional genomics for assessing the health impact of environmental contaminants and for bioremediation.
- Chemical processing, including
  — tank waste processing and separation systems for tank wastes as well as sensor design,
  — green chemical processing strategies for waste remediation, and chemical and petroleum production to minimize waste streams and energy consumption,
  — homogeneous and heterogeneous catalyst design (e.g., controlled oxidation of organics to produce intermediates for the chemical process industry or NO_x or SO_x emission reduction from combustion systems), and
  — models of the behavior of waste storage systems (e.g., glasses for radionuclide storage in Waste Isolation Pilot Plant or Yucca Mountain) over long time periods (hundreds of thousands of years).
- Developing a thorough understanding of combustion chemistry to reduce unwanted emissions (e.g., NO_x abatement strategies for lean-burn engines) and to improve system performance—a large number of chemical species and reactions are involved in the combustion of hydrocarbon fuels, and little is known about the highly reactive intermediates and many of the reactions.

Computational Design of Catalysts:
The Control of Chemical Transformation

The U.S. petroleum, chemical, biochemical, and pharmaceutical industries are the world's largest producer of chemicals, ranging from "wonder" drugs to paints to cosmetics to plastics to new more efficient energy sources. The U.S. chemical industry represents 10% of all U.S. manufacturing, employing more than one million Americans. It also is one of the few industries that has possessed a favorable balance of trade. The petroleum and chemical industries contribute ~$500 billion to the gross national product of the United States These industries rely for their financial well-being on their ability to produce new products by using energy-efficient, low-cost, environmentally clean processes, with a minimal number of undesirable side products. Key ingredients in 90% of chemical manufacturing processes are catalysts. A catalyst's role is to make a chemical reaction that produces a desired product proceed much more efficiently than it otherwise would by changing the kinetics of the process. Catalysis and catalytic processes account for nearly 20% of the U.S. gross domestic product and nearly 20% of all industrial products. Chemical transformations in industry take a cheap feedstock (usually some type of hydrocarbon) and convert it into a higher-value product by rearranging the carbon atoms and adding functional groups to the compound. About 5 quads per year are used in the production of the top 50 chemicals in the United States and catalytic routes account for the production of 30 of these chemicals, consuming 3 quads. Improved catalysts can increase efficiency leading to reduced energy requirements, while increasing product selectivity and concomitantly decreasing wastes and emissions. A process yield improvement of only 10% would save 0.23 quad per year! In addition, production of the top 50 chemicals leads to almost 21 billion pounds of CO_2 emitted to the atmosphere per year.[1] Improved catalysts can help reduce this carbon burden on the atmosphere. As new products become ever more sophisticated, the need to quickly develop new catalysts grows rapidly in importance. A fundamental understanding of chemical transformations is needed to enable scientists to address the *grand challenge of the precise control of molecular processes by using catalysts.*

Whereas Mother Nature is very effective at designing catalysts such as enzymes, we are decidedly less so. The most common approach to catalyst design used to be Edisonian. Try something; if it works, try to improve on the design by systematically changing the chemical nature of the catalyst; if it doesn't work, try something else. This approach is highly intensive, in terms of both time and expense, and most of the time, it did not work well. In addition, catalysts developed using this approach often produced undesirable by-products, and the catalyst itself may pose an environmental hazard. For many catalytic processes it is still unclear just how the catalyst works. A more desirable approach to catalyst design is to analyze at the molecular level exactly how catalysts function and to use this information to lead to the discovery of new systems and to optimize the design of others. Without this information, it is impossible to "tune" the catalyst to have the desired effect. For example, even the most sophisticated experimental techniques are unable to provide the details of the chemical reactions occurring at the surface of a heterogeneous catalyst or information about how to tune a homogeneous catalyst to gain a factor of 2 to 4 in performance.

The coupling of theory and experiment provides the most profound insights into catalyst behavior enabling the design of new catalysts. A combined theory-experiment approach was used to design a new alloy catalyst for ammonia synthesis leading to the first new ammonia catalyst since Haber and Bosch's work in the early 1900s that is better than iron. The development of a Lewis acidity scale based on high-level computations led directly to the design of a mixed metal catalyst for the liquid phase production of the hydrofluorocarbon refrigerant HFC-134a, the only such available liquid phase process.

The computational design of practical catalysts for industrial and commercial applications requires the ability to predict, at the molecular level, the detailed behavior of large, complex molecules as well as solid-state materials. Although intermediate-level computations can often provide insight into how a catalyst works, the true computational design of practical catalysts for industrial and commercial applications will require the ability to predict accurate thermodynamic

---

[1]Tonkovich, A. L.Y.; Gerber, M. A. *The Top 50 Commodity Chemicals: Impact of Catalytic Process Limitations on Energy, Environment and Economics,* PNL-10684, Pacific Northwest National Laboratory, Richland WA, 1995, 99352.

and kinetic results. There are enormous technical challenges in the computational design of catalysts. These include the number of different scales that must be considered, the size of the active domain, the need to treat heterogeneous structures, the need to consider the effective environment in which the catalyst acts (gas phase, solid phase, solution phase, at interfaces), and the need to treat complex metal interactions as most catalysts involve transition or lanthanide metals. Another set of challenges is that catalysts must be designed for use in real environments such as chemical reactors. The goals for computational catalysis on next-generation computer architectures are:

50 teraflops (10^9 floating-point operations per second): accurate calculations for realistic, isolated homogeneous catalyst model systems (<1.0 kcal/mol thermodynamics, <50% error in reaction rates).

250 teraflops: accurate calculations for realistic homogeneous catalyst model systems in solution and heterogeneous catalysts in vacuum (<1.0 kcal/mol thermodynamics, <50% error in reaction rates).

1000 teraflops: accurate calculations for realistic homogeneous catalyst model systems in solution and heterogeneous catalysts in solution (<1.0 kcal/mol thermodynamics, <50% error in reaction rates).

Success in these computational approaches will enable the long-sought goal in catalyst research of the design of efficient catalysts from first principles. Such new catalytic systems will revolutionize how we manufacture the chemicals that are such a part of everyday life, from efficient automotive fuels to polymers to cancer-fighting drugs, as well as the design of catalysts to minimize tropospheric environmental pollution from combustion energy systems. The ability to computationally design efficient new catalyst systems would lead to a revolution in the design of green chemical manufacturing processes that minimize (1) the usage of raw materials, (2) energy utilization, and (3) the environmental impact of the process and waste stream. Improved catalytic converters for combustion systems would enable the use of different fuel types as well as the development of lean-burn engines that minimize the impact of engine emissions such as NO_x on the formation of tropospheric pollutants such as ozone and harmful aerosols. Furthermore, the development of lean-burn engines with low emissions would have a large impact on reducing CO_2 emissions to the atmosphere because these engines burn the fuel much more efficiently.

Computational Environmental Molecular Science for DOE Site Cleanup

Computational molecular science is a key technology for addressing the complex environmental cleanup problems facing the Department of Energy's nuclear production sites as well as other polluted sites in the nation. Production of nuclear weapons at U.S. DOE facilities across the nation over four decades has resulted in the interim storage of millions of gallons of highly radioactive mixed wastes in hundreds of underground tanks, extensive contamination of the soil and groundwater at thousands of sites, and hundreds of buildings that must be decontaminated and decommissioned. The single most challenging environmental issue confronting the DOE, and perhaps the nation, is the safe and cost-effective management and remediation of these wastes. Environmental impacts at these sites range from minimal (e.g., near-surface contamination with uncontaminated groundwater aquifers) to extensive (e.g., surface, vadose zone, and groundwater contamination that extends off-site). The DOE invests approximately $6 billion per year in environmental cleanup activities at its former and present production sites. Even with this large expenditure expected over a 50-year time frame, it is difficult to see how remediation of the sites will be accomplished by using currently available technology. The sole use of conventional approaches to remediation and control (e.g., excavation, treatment, recovery, and disposal of residual waste for contaminated soils; pump and treat for contaminated aquifers) is cost prohibitive. Remediation strategies for DOE sites will require the combination of conventional remediation approaches with the increasing effectiveness and decreased cost of emerging technologies (e.g., in situ techniques such as natural bioremediation), to meet future remediation goals within budget constraints. Incorporation of in situ remediation technologies into overall remediation strategies has the potential for significant cost savings by leveraging physical, chemical, and biological subsurface processes to enhance the natural recovery of vadose zone and groundwater systems. However, a key ingredient in the success as well as the cost-effectiveness of remediation efforts is fundamental knowledge about the chemical properties and interactions of the wastes with their environment. This information is needed in order to develop and use in situ processes as unlike ex situ processes, there is usually no easy way to effectively halt an in situ process once it has begun. Computational molecular science in combination with experimental investigations provides this fundamental understanding of the complex interactions of (man-made) materials with the environment.

To support the development of innovative technologies for remediating various DOE sites, we need to develop reliable models to investigate the impact of the technology and the appropriate level of risk of using the technology or of doing nothing. These models of contaminant fate and transport in the subsurface have to be built on a detailed understanding of the binding and reaction of contaminants on soil particles as well as transport and reaction in groundwater. In addition, reliable models of the direct impact of mobile contaminants on humans and the risk of proposed remediation technologies will be critical for developing the safest and most cost-effective approaches to site cleanup and for public acceptance of the cleanup process and results. It is clear that such research spans the interfacial regime from bare solid surfaces to complex, solution-phase surface chemistry, and covers a range of time and length scales. Thus, one important focus of such a computational science effort is an understanding of the linkages between different temporal and spatial scales. No matter how many details of the physics are included in the transport models, if the critical underlying physical, chemical, and biological data describing the various reactions are missing or unreliable, the accurate predictive capability of such models will be significantly lessened. Computational molecular science can provide the thermodynamic, kinetic, and structural properties data needed for the models and can provide data that is difficult, or at times even impossible, to obtain in the laboratory or in the field due to the cost or danger of the experiment. The calculation of interactions of chemicals, including those containing radioactive elements, with environmental matrices such as soils is incredibly complex and will require tens to hundreds of sustained teraflops to begin to reliably predict the molecular interactions of chemicals with environmental systems and to provide the underlying data needed for reactive transport models. High-quality data are needed, and great care must be taken to minimize the errors in the calculated underlying data used in a sophisticated environmental or chemical process model so that errors in the data do not accumulate, propagate, and ultimately invalidate the macroscopic-scale model. There is a time-criticality in the need for the data that this computational molecular science effort will provide. Contaminants have been released into the environment over the past 50 years, and their mobility has led, and is continuing to lead, to broad-scale soil and groundwater contamination of the DOE sites.

Some of the most hazardous materials in the underground tanks, in the soils and groundwater, and in the buildings contain radioactive isotopes of the actinides and lanthanides. These materials pose significant radiation and other health hazards. A basic understanding of actinide and lanthanide chemistry is required to develop the new technologies needed for remediating the sites. Because of the difficulty and expense involved in conducting experiments with radioactive materials, it is important to employ computational methodologies that include the effects of Einstein's theory of relativity to support accurate calculations on molecular systems containing the actinides and lanthanides in order to guide the choice of experiments and to reliably extend the available experimental data into all of the regimes of interest. Such a capability would minimize the need for experimental work on radioactive materials. This capability would support many other research programs of importance to DOE, including the development of highly efficient and selective catalysts for new industrial processes. In addition, such a computational capability is needed to help offset the nation's loss of expertise in the area of actinide chemistry that is so critical to the multiple missions of the DOE. Computational molecular science also can be used to aid in the design of new processes, for example, the design of new compounds for efficient separation of radionuclides, such as technetium, or the rational design of enzymes to enhance the biodegradation of organic wastes or to immobilize radionuclides. Computational molecular science not only provides needed data but also can be used to provide new insights and answer "what if?" questions raised by scientists and engineers. Improving the response time to answer such questions will dramatically shorten the design and development time for new remediation strategies. In summary, theory can, at enormous but feasible computational expense, reliably and safely predict the chemical and physical properties of radioactive substances, often more cost-effectively than performing an experiment. The science goals in this area for next-generation computer architectures are:

50 TFlops: accurate calculations for realistic models of lanthanides and actinides on complex mineral surfaces (<1.0 kcal/mol thermodynamics) to develop parameters for reactive transport models of the vadose zone

250 TFlops: accurate calculations for realistic models of lanthanides and actinides on complex mineral surfaces interacting with aqueous solutions (<1.0 kcal/mol thermodynamics) to develop parameters for reactive transport models of the vadose zone

1000 TFlops: accurate calculations for realistic models of lanthanides and actinides on complex mineral surfaces interacting with aqueous solutions (<50% error in reaction rates) to develop parameters for reactive transport models of the vadose zone

Not surprisingly, the ability to predict the transport of wastes and associated contaminants within the Earth's shallow crust has become one of the most important challenges facing the DOE and is one of the most daunting scientific and computational challenges. Reliable prediction of the disposition of contaminants is critical for decisions on virtually every waste disposal option, from remediation technologies such as in situ bioremediation to evaluations of the safety of nuclear waste repositories. Nuclear waste repository designs, for both high-level and low-level wastes, require knowledge of transport of fluids in the vadose and saturated zones. Such predictive capabilities also will impact the responsible utilization of our nation's energy resources, particularly with respect to oil and gas. Any such predictions must be based on a comprehensive understanding of processes in the Earth's shallow crust. The accurate computation of the transport and disposition over long spatial and temporal scales of fluid-borne contaminants in complex natural systems is well beyond current capabilities. We currently lack the scientific understanding and the computational resources necessary to address these issues with the desired accuracy. However, with future generations of computational resources, it will be possible to address scientific and technical issues that are fundamental for quantitative analysis of fluid transport in terrestrial systems and essential for improving predictive capabilities.

Computation is essential for predicting the intricate web of species, reactions, and interconnections of transport in the natural system. The few natural examples of the integrated record of these processes constitute experiments that cannot be offered as an alternative to computational approaches. In addition, computational simulations are the only means that we have for predicting how contaminants will move over a long time period to predict the future state and the potential long-term impact of proposed remediation strategies. Computational modeling is the essential tool for translating scientific breakthroughs into practical applications in the area of subsurface transport. There are two key challenges for predictive modeling of subsurface transport. First, the present generation of hardware and software is insufficient to address the comprehensive treatment of flow and transport of multiple fluid and solid phases in heterogeneous, three-dimensional, nonequilibrium, interactive systems such as those found in the natural state. Second, as discussed above, accurate theoretical descriptions of the mechanisms of interaction are not yet fully developed. Overarching issues concern the ability to scale processes over many orders of magnitude in length and time, the ability to handle coupled complex relationships of processes at multiple scales, and the ability to predict and evaluate the results of processes over long time scales using data available only for narrow windows of observation. Over the next 10 years, simulation will play a major role in the development of new theory for field-scale simulations. A hierarchy of models designed for different scales is necessary to identify key features and processes at fundamental scales that have to be propagated to the field scale. Key scales where integration is needed include

1, the microbial membrane and surface where molecular-level structure and binding reactions are important to the biogeochemistry;

2. the mineral surface in solution to provide the setting for the interaction of minerals, aqueous components, surface-complexed metals, and attached microbial populations;

3. a network of pores to put multiple mineral surfaces in contact with transporting fluids that move nonuniformly, allowing for varying rates of exposure to nutrients and reacting components;

4. at the Darcy scale (macroscopic scale—for fluid volumes on the order of milliliters), similar degrees of process interaction are necessary, but the process models are based on bulk parameterizations to account for behaviors that cannot be resolved by the continuum approach; and

5. at the field scale (megascopic scale) process integration is similar in approach to the Darcy scale but generally considers more range in the parameter space (e.g., heterogeneity) but less detail in spatial and temporal resolution.

Subsurface simulations have actually begun to impact geochemistry, geophysics, and environmental chemistry by encouraging holistic and mechanistically detailed investigations where multiple processes and multiple reactive components are considered and monitored in complex subsurface materials. Over the next 10 years, these simulations are expected to be indispensable to the advancement of subsurface science through the systematic upscaling of processes at fundamental scales to representative parameterizations useful to field-scale models.

Atmospheric Chemistry

Problems in atmospheric chemistry are extremely challenging computationally. They are highly nonlinear, require the ability to simulate processes occurring on a vast range of spatial and temporal scales, and require coupled simulation of complex systems (the ocean and atmosphere) involving dynamics, thermodynamics (including condensable gases), chemistry, and thermal radiation. They are compute intensive, data-intensive, memory-intensive, and analysis-intensive. One area in which computational chemistry has played an important role in atmospheric chemistry is in the development of alternatives to the chlorofluorocarbons involved in stratospheric ozone depletion. Computational chemistry was essential for calculating missing thermodynamic data as well as correcting older experimental data on the CFCs, hydrochlorofluorocarbon (HCFCs), and hydrofluorocarbons (HFCs). Indeed, there are now far more calculated values being used than experimental ones. Calculations were critical to understanding the stability of an alternative in the environment and to the design of production processes and the actual refrigeration system. The calculated thermodynamic and kinetic properties have been used extensively in the design of chemical plants for producing CFC alternatives as well as by researchers who are developing new catalysts and alternatives. From calculations, it is now possible to reliably predict the thermodynamics and kinetics of the atmospheric degradation processes for the CFC replacements that are so critical to the design of environmentally safe alternatives. The ability to reliably predict rate constants for the initial reaction of an HFC or HCFC with a hydroxyl radical and the ability to then develop reliable degradation mechanism parameters was important in understanding and predicting the behavior of the alternatives in the troposphere and stratosphere in terms of ozone depletion potential, and together with calculated infrared intensities, a compound's global warming potential could be calculated. The computational chemistry results could then be used in much larger-scale atmospheric models involving fluid flow. DuPont used the results from computational chemistry to improve its process design for the manufacture of new fluorochemicals. By using thermodynamic and kinetic parameters calculated from first principles, the behavior of a pilot plant for the production of a CFC alternative was modeled correctly before actual operation began. Computational work was critical to getting the CFCs replacements to market as soon as possible, helping to save the stratospheric ozone layer. Computational chemistry is continuing to be used in many areas of atmospheric chemistry as we improve our knowledge of how the atmosphere behaves. Modeling aerosol formation, attachment of chemicals to aerosols, and reactions on aerosols are very difficult computational problems that are just beginning to be addressed. Atmospheric models are extremely complex especially when the correct physics (e.g., radiative transport and improved cloud parameterizations) is included, and they represent another computational grand challenge crossing many scales in space and time.

Summary

As discussed above, computational chemistry can play a key role in advancing the scientific enterprise. It can provide the data input for many larger, more complex models and provide us with unique insights into molecular behavior so that we can design and construct new molecules for specific tasks. Computational chemistry has become an established tool in the chemist's toolbox and is being used in broad areas of chemistry to replace experimental measurements and to provide us with improved understanding of molecular behavior. Computation will be the major tool that enables us to cross the many temporal and spatial scales that characterize environmental science.

CHEMICALLY RELATED R&D AT THE EPA'S OFFICE OF RESEARCH AND DEVELOPMENT

William H. Farland
U.S. Environmental Protection Agency

This presentation will provide background information on programs at the Office of Research and Development (ORD) at the U.S. Environmental Protection Agency. EPA's mission is broadly designed around protecting human health and safeguarding the natural environment—aims that require the application of chemistry, biology, epidemiology, physical sciences, and engineering in an integrated fashion. The role of ORD is to provide scientific foundations to assist the EPA's work by:

- conducting research and development to identify, understand, and solve current and future environmental problems;
- providing responsive technical support to EPA's programs and regions;
- collaborating with our scientific partners in academia and other agencies, state and tribal governments, private sector organizations, and nations; and
- exercising leadership in addressing emerging environmental issues and advancing the science and technology of risk assessment and risk management.

ORD is a relatively large organization, with three national laboratories and three national centers. In early November 2002, we announced a new center for homeland security research in Cincinnati, and EPA has the lead on drinking water infrastructure issues for the security of the nation's drinking water system. We also played a significant role in the decontamination of congressional buildings that suffered anthrax exposure in 2001. Overall, ORD has about 1900 employees in R&D, about 1200 of whom have graduate degrees in science.

In addition to its role in both health and ecological research, EPA also has responsibility for research in pollution prevention and new technology activities.

Although there are other national institutes and centers that work in this area, only EPA has a fully integrated, multidisciplinary, problem-directed research program. We rely on the peer review process and the scientific community to focus our program as we push the envelope for being state of the art. At the same time we have a responsibility to evaluate the broader scientific literature and collect additional information to ensure that science is credibly used in EPA's decisions. This is accomplished through a combination of our in-house research program and an external grants program. Sound science is essential for EPA decisions:

• Science is a critical component of credible decisions and actions that protect human health and the environment.
• Making EPA decisions with sound science requires relevant, high-quality, cutting-edge research in human health, ecology, pollution control and prevention, and socioeconomics; proper characterization of scientific findings; and appropriate use of science in the decision process.
• ORD is a leader in environmental research, focusing its efforts and resources on those areas in which EPA can add the most value to reducing uncertainty in risk assessments and enhancing environmental risk management.

I appreciated the discussion during the workshop session on the MTBE issue. A decade ago at EPA, we were laying out research needs for oxygenates, and we were concerned . We strongly suspected that MTBE would be a problem in groundwater. The discussion at this workshop shows that the concern is still with us. This clearly illustrates the importance of life-cycle analysis when we evaluate substances that are introduced widely into the environment.

Many of the decisions made by EPA are risk based and have significant uncertainties associated with them. The only way we can make better decisions is to fill the knowledge gaps that lead to the uncertainties. For the scientific basis that we provide for regulatory decision making, the research falls into two general categories, problem-driven research and core research, where the former includes topics such as

• fine particulates in air,
• drinking water contaminants,
• diesel engine emissions, and
• mercury in air and water.

The core research includes topics such as

• ecological monitoring and assessment of ecological resources,
• health risks to sensitive populations,
• pollution prevention and green chemistry, and
• environmental economics.

These topics illustrate the need to bring a multidisciplinary approach to problem-driven research. The core research program is closer to what might be called basic research, and it provides a forward-looking approach to dealing with ecological issues, health risks, pollution prevention, and support for environmental economics that eventually will allow us to anticipate new problems at an early stage. The interaction between problem-driven and core research is reciprocal and iterative.

Chemistry will play a large role in the research areas of our own strategic plan. For example, work on particulate matter will allow us to understand the nature of the particles and their behavior in the atmosphere, develop the modeling that will predict their fate and transport along with the resulting human exposure, and understand the transition from exposure to dose that will enable health assessment work.

Drinking water provides another example in which the by-products of disinfection illustrate ways that environmental chemistry is part of our everyday activities. We have also been looking at the nature of the toxin associated with *Pfiesteria* in order to understand the chemistry by which *Pfiesteria* produces lesions in fish and potentially has the ability to poison humans. This is a very difficult chemistry problem. Similarly, endocrine disrupters have been discussed in this workshop, and we need to understand their mechanism of action and the structure-activity relationships that would allow us to make a judgment about interaction with specific types of cellular receptors. Clearly, research in pollution prevention and new technologies are heavily based in the chemical sciences.

A more focused list of ORD research interests includes a large number of topics that came up in today's workshop discussions:

- Particulate matter
- Drinking water
- Global change
- Endocrine disruptors
- Ecological risk
- Human health
- Pollution prevention and new technologies

As one example of endocrine disrupter work, EPA has been involved in dioxin reassessment for the last 11 years. This is an incredibly complex issue, with a huge amount of available information—from the chemistry of toxicity equivalents and understanding how dioxin-like congeners can play into the total load in the environment to an understanding of how dioxin works to affect cells essentially as an environmental hormone.

Persistent biocumulative toxic chemicals were discussed in the workshop as one of the top opportunities for advances in chemistry—understanding that these hydrophobic chemicals have the ability to persist and move around in the environment and are particularly difficult for us to control and remediate. This is another of the areas in which we work on a day-to-day basis, and again chemistry plays a large role.

In atmospheric chemistry, a particular focus of our R&D activities is the development of air quality models and scaling to models that can predict local, regional, and global impacts on air quality—from sources that we have the ability to control. A new generation of air-quality models looks at community modeling of air quality (CMAQ), and this has provided a real advance for some of the regulatory activities that go on in EPA's program. Mercury has recently been added to the model, and certainly the biggest improvement will be addition of the NO_x module, which is currently under development. This will be particularly important if we move forward with a multipollutant approach to air pollution using the Clear Skies initiative.[1] Understanding the interactions among mercury, NO_x, SO_x, and other chemicals will be particularly important for us. ORD has supported work in Riverside, California to build chambers in which chemical reactivity studies will provide input into these models and advance our understanding of this issue.

Our programs include several areas that have a focus on analytical chemistry. For example, the Clean Air Act requires us to publish Federal Reference Methods for how one measures particulate matter. We have the responsibility for looking at methods that are functionally equivalent to the Federal Reference Methods so we have to be in a position to look at new analytical approaches. As another example, genomics has become an important way to look at molds as indoor air pollutants or microbial contamination on beaches. We need to develop markers that can be used for rapid detection and characterization of beach contamination—rather than growing cultures that might tell you tomorrow where you shouldn't have gone swimming yesterday. These applications combine genomics and chemistry, and they have important implications for children's health, and for determining the ability to inhabit residences after flooding or water damage from fires. This approach will allow us to identify the organisms, understand the remediation techniques that must be used, and evaluate the extent of cleanup. It is an important contribution from our laboratories.

Another analytical topic of concern to us is continuous monitoring. We're interested in finding ways of continuous monitoring for dioxins, mercury, and combustion by-products. Some of this work is also taking place in our laboratory in Las Vegas using pattern recognition from a continuous gas chromatography–mass spectrometry (GC-MS) analysis of water sources to discover changes, characterize the materials responsible for the change, and ultimately, respond quickly by diverting them from drinking water.

We have several hundred chemists on our research and development staff in-

---

[1]*http://www.epa.gov/clearskies/*

house primarily in our National Exposure Research Laboratory and our National Risk Management Laboratory. These individuals are, for the most part, research chemists who work on the kinds of problems I have described.

In 1995 we developed an approach to reaching out to the scientific community and investing funds in particularly important problems where we can engage some of the top scientists in the country. About a quarter of the academic faculty involved in this meeting are currently funded by our Science To Achieve Results (STAR) program, which provides about $100 million a year in support of basic research directed toward specific problems. STAR-funded research supports a broad range of issues that will have important payoffs for our future approaches to dealing with environmental problems. Some of the STAR solicitations that have a chemistry emphasis include:

- Measurement, Modeling and Analysis Methods for Airborne Carbonaceous Fine PM (2003)
- Development of High-Throughput Screening Approaches for Prioritizing Chemical for the EDC Screening Program (2003)
- Technology for a Sustainable Environment (National Science Foundation) (2001, 2003)
- Assessing the Consequences of Global Change for Air Quality: Sensitivity of U.S. Air Quality to Climate Change and Future Global Impacts (2002)
- Nutrient Science for Improved Watershed Management Program (2002)
- Environmental Futures Research in Nanoscale Science, Engineering and Technology (2001, 2002)
- Mercury: Transport, Transformation, and Fate in the Atmosphere (2001)

In a collaborative program called Technologies for a Sustainable Environment, the National Science Foundation (NSF) and EPA have awarded approximately $46 million for 164 projects addressing

- environmentally benign solvents,
- biotechnology for pollution prevention,
- green chemistry-reaction modifications,
- green engineering-process modifications, and
- industrial ecology-environmentally benign manufacturing.

Pollution-prevention research is an overarching activity for which we want a combined in-house and extramural program that is focused on finding ways to eliminate or minimize hazardous solvents, emissions, and waste. In other words, we want to clean up processes to reduce impacts on the environment; to examine renewable feedstocks as a process improvement; and to minimize water, energy, and materials use in industrial processes. The work that EPA and NSF have funded on pollution prevention has been dominated by the chemical engineering

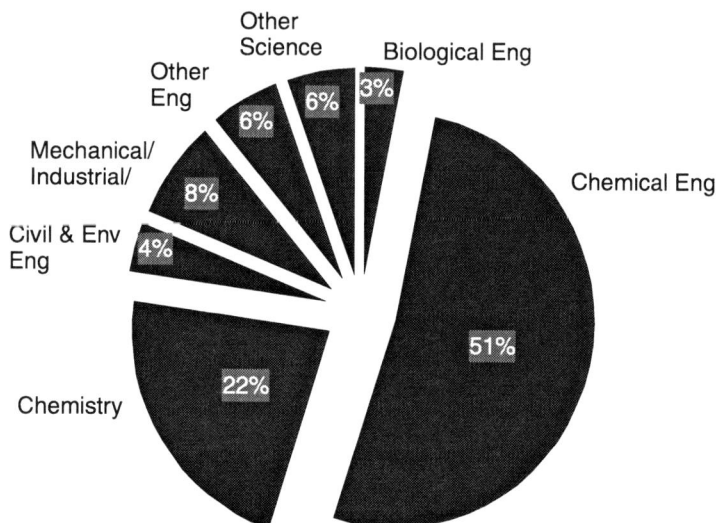

FIGURE 1 Chemical sciences dominate pollution-prevention grants portfolio.

and chemistry community (Figure 1). We hope that there will be a continued commitment in the areas of pollution prevention, greener chemistry, and greener technologies. I think this is a success story that illustrates how problem-directed research in these disciplines will be important to us for the future.

I agree with many of the workshop participants that we have an opportunity in terms of nanotechnology. Certainly there are indications that we can improve environmental sensing, treatment, and remediation if we can improve manufacturing processing efficiency, reduce waste production and toxicity, and reduce materials consumption. What are the effects of nanomaterials on the environment?

• Currently, little is known about the potential effects of manufactured nanoparticles on human health and the environment.
• Nanomaterials may enter the food chain and human body when released into the air or water or discarded on the land.
  • Nanomaterials could affect human health and the environment through
    — Exposure to skin,
    — Adsorption by the lungs, and
    — Bio-uptake and bioaccumulation.
• The toxicity of nanomaterials is largely unknown.

All of this is chemistry and chemical engineering at its finest if we can find a way to make it work at a nanoscale level. This is the real challenge for us. We participated in an interagency National Nanotechnology Initiative that will invest about $710 million in this particular area this year. The vast majority of this work comes out of NSF, the Department of Defense (DOD), and DOE. But EPA invests about $5 million a year right now in this area. We are making an important contribution in understanding what needs to be done in integrating the issues with environmental problems.

In summary, ORD has a dynamic program of research and development that really is based on integration of disciplines. We are dealing with problems for which solutions necessitate an integration of various science and engineering disciplines. We have a balance between problem-driven and core or basic research that requires us to really build a strong relationship with our partners, both inside the agency and out in the general scientific community, to meet the agency's needs, and we hope to be able to continue to enhance that partnership. We are in a position to anticipate some of the emerging scientific issues and, hopefully, be poised to meet the challenges of the twenty-first century.

BIOGEOCHEMICAL CONTROLS ON THE OCCURRENCE AND MOBILITY OF TRACE METALS IN GROUNDWATER

Janet G. Hering
California Institute of Technology

Introduction

Clean water has been called the "oil of twenty-first century," a phrase that reflects both the growing demand for water resources and the recognition that the quality of many water resources has been degraded by human activities. The importance of groundwater as a water supply in developed countries, such as the United States, is often overlooked, yet in 1995, 46% of the domestic water supply was provided by groundwater and 54% by surface water (Table 1). Although groundwater is particularly important for self-supply in rural areas, it is also an important resource for public water systems. Demand for and dependence on groundwater supplies are expected to increase with increasing population in the United States, particularly in the West.

There are some important differences between groundwater and surface water with respect to water quality. Surface water is considerably more vulnerable to pathogens; groundwater provides some level of "natural protection" against bacteria and viruses that can cause outbreaks of human disease. The composition of groundwater, however, is more influenced than that of surface water by contact with soil and aquifer minerals, which generally leads to the accumulation in groundwater of constituents derived from geologic materials.

TABLE 1 Use of Surface Water vs. Groundwater for U.S. Domestic Supply in 1995

	Million Gallons per day (MGD)	Million Cubic Meters per Day (m³/d)
Surface water		
Self-supply	38	0.14
Public supply	14,100	53.4
Total	14,100	53.4
Groundwater		
Self-supply	3,350	12.7
Public supply	8,460	32.0
Total	11,800	44.7

SOURCE: Data from *http://water.usgs.gov/watuse/pdf1995/html.*

The natural occurrence of arsenic at elevated concentrations in groundwater in West Bengal, India, Bangladesh, and other parts of South Asia provides an unfortunate example of the potentially devastating effects of groundwater quality on human health.[1] In these areas, huge shifts in water resource utilization, from surface water to groundwater, have occurred over the past 30 years. Now tens of millions of tubewells providing drinking water for individual families and larger wells are used for groundwater-based irrigation. Arsenic occurs in the groundwater in these regions at concentrations up to the milligram-per-liter range, and concentrations between 200 and 800 µg/L are common. The human health effects of consuming this arsenic-contaminated water range from skin lesions to fatal cancers. At the same time, it must be recognized that groundwater offers protection from pathogens—diarrheal diseases are a major cause of death in infants in the developing world. Furthermore, groundwater-based irrigation has allowed substantial expansion of agricultural activities and has drastically improved the nutritional status of people in these areas. Nonetheless, the lack of attention to groundwater quality has had profound human health consequences, and these experiences illustrate the importance of evaluating the quality as well as the quantity of groundwater resources.

Properties of Aquifers Influencing Groundwater Quality

If we are to consider how the chemical composition of water withdrawn from a specific well has evolved, a key issue is the origin of the water and the contact

---

[1]Smith, A. H.; Lingas, E. O.; Rahman. M. *Contamination of drinking-water by arsenic in Bangladesh: A public health emergency. Bull. World Health Organ.* **2000**, *78(9)*:1093-1103.

it has had with soil and aquifer minerals between the sites of recharge and withdrawal.[2] A distinction should therefore be made between wells drilled into unconfined and confined aquifers. In unconfined aquifers, the water table is in contact with the unsaturated zone and is subject to local recharge; groundwater ages can be less than decades. In confined aquifers, however, a confining layer (usually clay) prevents local recharge, which must therefore occur in upland areas that can be distant from the withdrawal site. Residence times can vary greatly and groundwaters can be tens of thousands of years old.

Within an aquifer, water flows in response to the hydraulic gradient (i.e., from locations of higher to lower hydraulic head). However, flow velocities are also determined by properties of the aquifer materials, specifically hydraulic conductivity and porosity, as shown in the following equation:

$$\text{Average linear velocity} = \frac{K}{\eta}\frac{dh}{dL}$$

where K = hydraulic conductivity, $\frac{dh}{dL}$ = hydraulic gradient, and η = porosity.

For various aquifer minerals, porosity varies over a fairly narrow range (ca. 0.3 to 0.5) but hydraulic conductivity varies over many orders of magnitude.[2] Even for a specific type of aquifer material, ranges of 1-4 orders of magnitude are common (e.g., $10^{-8.5}$ to 10^{-4} m/s for fractured rock, 10^{-5} to 10^{-3} m/s for well-sorted sand). The lowest hydraulic conductivities are found for crystalline rock (10^{-14} to 10^{-10} m/s) and the highest for well-sorted gravel (10^{-2} to 1 m/s) and clean sand or cavernous limestone (10^{-6} to 10^{-2} m/s).

The prediction of groundwater flow is complicated by the heterogeneities of the subsurface environment, which occur on multiple scales. To understand the evolution of groundwater composition, it is also important to distinguish between flow through porous media and flow through channels (caused by dissolution of the rock matrix) or fractures (resulting from tectonic activity). Such differences in texture can result in differences in the contact between the fluid and mineral surfaces.

Biogeochemical Processes Affecting Groundwater Composition

In the physical context of water movement in subsurface and water-rock contact, we may then consider the biogeochemical processes that can affect the distribution of a constituent X between the immobile and mobile phases in the subsurface (Table 2). Obviously, mobile constituents are of the most direct concern because of the potential for human exposure. Constituents in the mobile

---

[2]Langmuir, D. *Aqueous Environmental Geochemistry;* Prentice Hall: Upper Saddle River, NJ **1997**.

TABLE 2 Partitioning of Constituent X Between Immobile and Mobile Phases

Immobile Phase	Process[a]	Mobile Phase
X as component of aquifer mineral matrix	Dissolution → ← Precipitation	Dissolved X
X sorbed on surfaces of aquifer materials	Desorption → ← Adsorption	Dissolved X
X dissolved in immobile water trapped in micropores	← Diffusion→	Dissolved X
X sorbed-precipitated in attached colloids	Detachment-peptization → ← Attachment-filtration	X sorbed-precipitated in mobile colloids

[a]Both precipitation-dissolution and sorption-desorption can be strongly influenced by microbial processes. Microbial redox transformations of constituent X may (especially for metals) significantly alter solubility, and microbial production of ligands can promote the release of X from immobile phases by dissolution and desorption and stabilize X in solution.

phase (i.e., as dissolved or colloidal species) can migrate from their source region to a drinking water well. However, it must be remembered that immobile constituents may have the potential to be mobilized if conditions in the subsurface environment change.

As indicated in Table 2, the type of process by which constituent X may be mobilized (or sequestered) depends on speciation of X in the solid phase. Thus, if X is an integral constituent of an aquifer mineral, dissolution of the mineral will be required to release X into solution, whereas mineral dissolution would not be required to mobilize X if it is sorbed onto a mineral surface. Environmental conditions (including both chemical composition of pore fluids and microbial activity) will influence the extent of mobilization or sequestration of X. For metals, in particular, both the solubility and the affinity for surfaces can be strongly influenced by redox conditions and the presence of (biogenic) complexing agents.

In order to gain insight into the evolution of groundwater composition, these biogeochemical processes must be examined, and subsurface materials characterized, over multiple scales ranging from nanometers to kilometers (Figure 1). Each level of investigation in Figure 1 corresponds to different types of process investigation or material characterization. For example, at the nanoscale, x-ray absorption spectroscopy (XAS) has been a powerful tool to characterize the local (i.e., coordination) environments of a range of elements in association with solid surfaces.[3] At the laboratory scale, macroscopic experiments have determined disso-

---

[3] Brown, G. E.; Parks, G. A. *Int. Geol. Rev.* **1992**, *43(11)*, 963-1073.

```
┌─────────────────────────┐
│          Field          │
│      Observations       │
└─────────────────────────┘
              │
┌─────────────────────────┐
│   Field or "Mesocosm"   │
│      Experiments        │
└─────────────────────────┘
              │
┌─────────────────────────┐
│       Laboratory        │
│      Experiments        │
└─────────────────────────┘
              │
┌─────────────────────────┐
│        Nanoscale        │
│    Characterization     │
└─────────────────────────┘
```

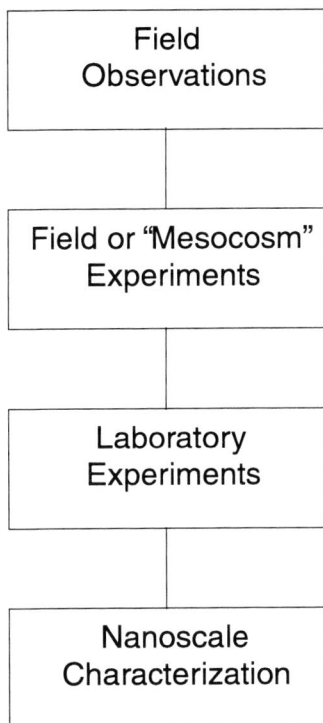

FIGURE 1 Interrogation of multiscale subsurface biogeochemical processes.

lution rates for a wide variety of minerals as well as their properties as sorbents for both inorganic and organic chemical species.[4] The effects of factors such as solution composition, the type and surface area of the solid, and biological activity on both dissolution and sorption processes have been studied at this scale. Field- or mesoscale experiments include investigations conducted at the U.S. Geological Survey (USGS) Cape Cod Toxic Substances Hydrology Research Site, which is intensively instrumented with multilevel samplers and where subsurface conditions can be manipulated by injection of hundreds of liters of water with modified composition.[5] Similar manipulations have been performed as a "push-pull" experiment at an individual well.[6] In field observations, evolution of

---

[4]Stumm, W. *Chemistry of the Solid-Water Interface*; Wiley-Interscience: New York, 1992.

[5]USGS. Cape Cod Toxic Substances Hydrology Research Site. U.S. Geological Survey, 2002; *http://ma.water.usgs.gov/CapeCodToxics/*.

[6] Harvey, C. F.; Swartz, C. H.; Badruzzaman, A. B. M.; Keon-Blute, N.; Yu, W.; Ali, M. A.; Jay, J.; Beckie, R.; Niedan, V.; Brabander, D.; Oates, P. M.; Ashfaque, K. N.; Islam, S.; Hemond, H. F.; Ahmed. M. F. *Science* **2002,** *298(5598),* 1602-1606.

groundwater quality can be related to aquifer characteristics and attributed to processes occurring within the aquifer.[7]

Linking these different scales presents a substantial challenge. A critical issue is to determine how information acquired at the nano- or molecular scale can be applied to elucidate macroscale processes in the environment. Modeling and simulation provide an avenue for the linkage of biogeochemical processes across multiple scales. Ab initio calculations and molecular dynamics simulations are powerful tools for linking nano- and molecular-scale observations to macroscopic behavior on the laboratory scale.[8] A variety of transport codes have been developed to model flow in the subsurface. Although the level of biogeochemical sophistication incorporated in these codes varies widely, a few do incorporate biogeochemical kinetics as well as equilibrium reactions. The USGS code PHREEQC[9] combines reaction kinetics with one-dimensional transport, and the code HydroBioGeoChem 123D includes both three-dimensional flow and transport in the vadose (i.e., unsaturated) zone.[10] A key issue is to understand what information about the subsurface environment and the operative biogeochemical processes is needed to constrain calculations made using these codes so that they have real predictive value.

Case Study: Arsenic Mobilization in Sediments

As already mentioned, the mobilization of arsenic provides an example of the enormous impact that biogeochemical processes can have on groundwater quality and human health. In the Ganges-Brahmaputra delta (i.e., Bangladesh and West Bengal, India), arsenic mobilization has been attributed to the reductive dissolution of iron oxides and the release of arsenic associated with these carrier phases.[11] Recent manipulation experiments have shown enhanced release of arsenic with the injection of molasses (as an organic substrate for microbial iron reduction) and arsenic sequestration with the injection of nitrate (as an electron acceptor for microbial iron oxidation).[6] Possible sources of organic matter whose oxidation supports iron reduction and the elevated arsenic concentrations ob-

[7]Plummer, L. N.; Busby, J. F.; Lee, R. W.; Hanshaw, B. B. *Water Resour. Res.* **1990,** *26(9),* 1981-2014.

[8]*Molecular Modeling Theory : Applications in the Geosciences;* Cygan, R. T.; Kubicki, J. D., Eds.; Reviews in Mineralogy and Geochemistry, vol. 42; Mineralogical Society of America: Washington, DC, 2001.

[9]USGS. PHREEQC (version 2): a computer program for speciation, batch-reaction, one-dimensional transport, and inverse geochemical calculations. U.S. Geological Survey, **2002;** *http://wwwbrr.cr.usgs.gov/projects/GWC_coupled/phreeqc/index.html.*

[10]Gwo, J. P.; Frenzel, H.; D'Azevedo, E.; Hoffman, F. M. *Hydrobiogeochem.* **1999,** 123D, v. 1.1, Oak Ridge National Laboratory.

[11]Nickson, R.; McArthur, J.; Burgess, W.; Ahmed, K. M.; Ravenscroft, P.; Rahman, M. *Nature* **1998,** *395,* 338.

served in these aquifers include buried peat deposits[12,13] and newly infiltrated organic carbon derived from agricultural activities.[6]

The coupled biogeochemical cycling of iron and arsenic has also recently been studied in a system dominated by the addition of iron as an engineering practice.[14] Since 1996, ferric chloride has been added as a coagulant to the Los Angeles Aqueduct in order to control the levels of naturally occurring arsenic that reach the water distribution system for the City of Los Angeles. The iron- and arsenic-rich floc formed by this process is deposited in North Haiwee Reservoir. Sediments and porewaters from this reservoir were examined to determine the extent to which arsenic and iron are remobilized in the sediments and to probe the speciation of arsenic in the solid phase and its possible effects on arsenic remobilization.

Arsenic and iron concentrations in the sediment porewaters were found to be closely correlated. Although these concentrations are considerably elevated at depth in the sediment column—reaching 17 µM (1.3 mg/L) for arsenic and 1.6 mM (90 mg/L) for iron—only a small fraction of the iron and arsenic deposited to the sediments needs to be remobilized to support these concentrations. XAS analysis of the sediments indicated that arsenic in the solid phase is reduced from As(V) to As(III) above the depth at which arsenic is released into the porewater. Iron in solid phase remains as Fe(III). XAS analysis showed no evidence of conversion to magnetite (though conversion of ferrihydrite to goethite could not be excluded). Sequential extractions indicated that most of the arsenic can be released from sediment by treatment with magnesium chloride or phosphate solutions; this treatment does not release iron, behavior that is consistent with sorption as a mode of association for the majority of the arsenic with the sediment. The remainder of the arsenic is released, along with almost all the iron, by treatment with hydrochloric acid.[15]

These observations indicate that reduction of As(V) to As(III) does not, in itself, result in the mobilization of arsenic. This conclusion is supported by laboratory adsorption studies showing similar affinities of As(III) and As(V) for hydrous ferric oxide, goethite, and magnetite.[16] However, outstanding questions remain regarding the factors that control the rate and extent of the reductive dissolution of iron in these sediments and whether the arsenic (and iron) that is released into the porewater is (re)sorbed onto the residual iron oxyhydroxides in

[12]Nickson, R. T.; McArthur, J. M.; Ravenscroft, P.; Burgess, W. G.; Ahmed, K. M. *Appl. Geochem.* **2000,** *15,* 403-413.

[13]McArthur, J. M.; Ravenscroft, P.; Safiulla, S.; Thirlwall, M. F. *Water Resour. Res.* **2001,** *37(1),* 109-117.

[14]Kneebone, P. E.; O'Day, P. A.; Jones, N.; Hering J. G. *Environ. Sci. Technol.* **2002,** *36,* 381-386.

[15]Dempsey, D. *Arsenic Cycling in Natural Sediments: Effects of Aging,* SURF report: California Institute of Technology; Pasadena, CA, 2002.

[16]Dixit, S.; Hering, J. G. *Environ. Sci. Technol.,* **2003,** submitted.

the sediment. These are the types of questions that must be answered in order to predict the concentration of arsenic in the porewater and its response to possible changes in environmental conditions.

Opportunities, Challenges, and Needs

Groundwater is an important resource and one that will be critical in meeting future needs for water for human consumption, irrigation, and industrial uses. Groundwater quality is an important (though sometimes underappreciated) issue because naturally occurring groundwater constituents can adversely affect human health. Although arsenic has been highlighted here, other trace constituents of geologic materials, such as uranium and radon, also occur naturally in groundwater and have known adverse health effects. Chromium(VI) has recently been identified as a natural constituent of groundwater in California though its health effects are less well understood. Aesthetic problems are posed by some common groundwater constituents such as iron, manganese, and sulfide.

In addition, groundwater quality can be degraded by introduction of contaminants from a wide range of human activities. These include increased salinity, nitrate, pesticides, and pathogens from agriculture; pathogens from septic systems; fuel hydrocarbons and oxygenates (e.g., MTBE) from underground storage tanks; chlorinated solvents, metals, and perchlorate from industrial manufacturing; metals from mining, ore processing, and refining; and radionuclides from nuclear weapons and energy production. Because of the long residence time of groundwater and the inaccessibility and/or physical extent of sources of contamination in the subsurface, groundwater remediation is often a slow and difficult process.

In order to use and manage groundwater resources productively and safely, it is necessary to understand the occurrence and mobility of naturally occurring groundwater constituents and the fate and transport of contaminants in the subsurface environment. Insight into the biogeochemical processes by which chemical species are mobilized from or sequestered into immobile phases in the subsurface is crucial to this understanding.

The chemical sciences offer insight into the molecular mechanisms of biogeochemical processes, links between nano-, laboratory, and field scales through modeling and simulation, sensors for the detection and quantification of chemical constituents of groundwater, and chemical reagents and chemistry-based technologies for in situ remediation of contaminated aquifers.

Yet, various needs must be addressed if the challenge of providing safe and adequate water supplies is to be met. The interrogation of biogeochemical processes and subsurface materials at a fundamental level requires continued support that recognizes both the value of studying well-controlled model systems and the need to work directly in complex, environmental systems. Multidisciplinary collaborations that examine complex systems must be supported and complemented by focused, disciplinary research. Priority should be given to the development and application of sensors for use in the subsurface environment and of tools for the characterization of subsurface materials at the nano- and atomic scale. These latter efforts should support activities at shared or user facilities such as the Stanford Synchrotron Radiation Laboratory (*http://www-ssrl.slac.stanford.edu/*) and the Environmental Molecular Sciences Laboratory (*http://www.emsl.pnl.gov:2080/*). Finally, a crucial resource is provided by facilities, such as the USGS Cape Cod Toxic Substances Hydrology Research Site,[5] that allow field-scale experiments to be conducted. Support should be provided for activities at this and additional facilities, such as the Vadose Zone Research Park and the proposed Subsurface Geosciences Laboratory at the Idaho National Engineering and Environmental Laboratory (*http://www.inel.gov/env-energyscience/geo/*).

The occurrence and mobility of harmful chemical substances, whether of natural origin or anthropogenic contaminants, in the subsurface environment pose both an intellectual and fundamental scientific challenge and practical concerns for the use and management of groundwater resources. The chemical sciences offer powerful approaches toward understanding and mitigating the problems of groundwater contamination. Society has benefited and will continue to benefit from this important application of chemistry to environmental problems.

Acknowledgments

The collaboration of P. O'Day, S. Dixit, P. Kneebone, and D. Dempsey and the cooperation of the Los Angeles Department of Water and Power in projects described in this paper are gratefully acknowledged.

MEASUREMENT CHALLENGES AND STRATEGIES IN ATMOSPHERIC AND ENVIRONMENTAL CHEMISTRY

Charles E. Kolb
Aerodyne Research, Inc.

Atmospheric and environmental chemistry are rapidly evolving disciplines that play a critical role in a wide variety of current environmental issues spanning local, regional, continental, and global scales. Effective management of these issues generally requires knowledge of chemical and physical properties over a wide range of spatial and temporal scales, with processes involving fluid media (atmospheric, oceanic, surface, and groundwater) usually demanding frequent updating (seconds to months, depending on chemical lifetimes and/or transport rates). Chemical characterization methods that rely on traditional sample collection and subsequent analysis are generally slow, labor intensive, time consuming, and costly. Alternatively, environmental systems may be monitored pseudo-continuously at fixed-site instrumented stations, reporting ambient chemical species concentrations with time resolutions ranging from minutes to months. However, cost restraints usually severely limit the number of fixed-site monitoring stations deployed so that impacted environments are sparsely sampled spatially. The constraints imposed by traditional environmental measurement methods generally lead to environmental systems being badly undersampled in the spatial and/or temporal domains.

The undersampling of environmental systems often has several unfortunate consequences. First, it can lead to misdiagnosis of both the nature and extent of environmental problems, leading to either an over- or underestimation of the seriousness or extent of the problem. Second, environmental process models developed to assess the effectiveness of regulatory or other environmental management challenges are often "validated" with insufficient experimental data to truly constrain the model, resulting in potentially unproductive or even counterproductive management strategies. Third, misunderstanding the nature and/or extent of an environmental problem can obscure its interconnections with other environmental problems, creating the possibility that management strategies may help eliminate or mitigate the target issue but exacerbate connected problems.

Environmental measurements are challenging for a number of reasons that are summarized in the following section. After presenting measurement challenges, some promising strategies for addressing environmental management challenges are advanced. Atmospheric chemistry measurement challenges and strategies are emphasized, but many of the lessons learned there can be applied to environmental problems with water, soil, and ecological components.

Environmental Measurement Challenges

The world is a pretty large place. The Earth's surface area is ~500 million square kilometers, approximately two-thirds ocean and one-third land. For atmospheric issues we are typically concerned with the two lowest atmospheric regions, the troposphere and the stratosphere, which compose the first 50 km of the atmosphere, with a volume of ~25 billion cubic kilometers. So atmospheric chemical issues that are global in scope pose a real measurement challenge. Furthermore, as noted above, the atmosphere is dynamic, so chemical and physical characterization measurements must be repeated, sometime quite frequently, to well describe and predict the evolution of processes of interest. Also, because the atmosphere is highly dynamic, quantification of fluxes—surface-atmosphere emissions and depositions, stratospheric-tropospheric exchange, boundary layer detrainment, transport across the Intertropical Convergence Zone (ITCZ)—are often as or more important than ambient concentration measurements.

Atmospheric measurements are also challenging because they must deal with low to extremely low concentrations of trace chemical species. The major components (>99.999%) of the lowest portions of the atmosphere (the troposphere up to ~10 km in altitude and the stratosphere between ~10 and ~50 km) are molecular nitrogen, molecular oxygen, argon, water vapor, and carbon dioxide. Chemists will recognize that all of these species are very stable, strongly bonded molecules or atoms that are essentially inert gases at normal atmospheric temperatures (190-310 K). Indeed, without solar photons to break up selected molecules, atmospheric chemistry would be very dull indeed. Atmospheric chemistry is dominated by trace species, ranging in mixing ratios (mole fractions) from a few parts per million, for methane in the troposphere and ozone in the stratosphere, to hundredths of parts per trillion, or less, for highly reactive species such as the hydroxyl radical. It is also surprising that atmospheric condensed-phase material plays very important roles in atmospheric chemistry, since there is relatively so little of it. Atmospheric condensed-phase volume to gas-phase volume ratios range from about 3×10^{-7} for tropospheric clouds to ~3×10^{-14} for background stratospheric sulfate aerosol.

Progress in understanding key atmospheric environmental issues well enough to identify and test effective management strategies is very dependent on our ability to measure the required range of chemical and physical parameters over adequate spatial scales and appropriate time scales. It has been recognized for more than a decade that the inability of available instrumentation to meet these needs is a serious issue in both atmospheric chemistry research[1] and assessments of the effectiveness of air quality regulations.[2]

Environmental Measurement Strategies

It is important to recognize that environmental scientists make measurements for a variety of reasons, each of which imposes its own requirements and constraints on the instrumentation and measurement systems to be used. General environmental measurement modes include exploratory mapping and surveying, process investigations, baseline establishment and trend monitoring, and emission-deposition and other flux measurements. Table 1 lists some of the goals that drive each mode of measurement and some of the measurement system capabilities that they require.

While the various environmental measurement modes have a range of requirements, the undersampling problem discussed above is endemic. We are frequently unable to make enough measurements under an adequate range of environmental conditions over a sufficient range of spatial dimensions and time scales to understand and describe the issue being investigated. Three general strategies have been identified that can address the environmental undersampling issue.

[1] Albrittton, D. L.; Fehsenfeld, F. C. Tuck, A. F. *Science* **1990,** *258,* 75-81.

[2]*Rethinking the Ozone Problem in Urban and Regional Air Pollution,* National Research Council, National Academy Press, Washington, DC, 1991.

TABLE 1 Environmental Measurement Modes

Mode	Goal(s)	Typical Requirements
Exploratory mapping-surveying	Establish ranges of environmental variables Investigate spatial-temporal variability Check for surprises	Easy sampling-analysis, mobility
Process investigation	Quantitative understanding Predictive model	Multispecies-multiparameter measurements, high temporal and spatial resolution, flux measurements
Baseline-trend monitoring	Quantify current state and rate of change Determine if regulations-management strategies are working	High precision, regular repetition, long-term stability, adequate spatial and temporal resolution and coverage
Emission-deposition measurements	Establish pollutant sources and quantify fluxes Check estimation algorithms Improve environment models Determine if regulations-management strategies are working	Mobility, moderate to very high temporal and spatial resolution, ability to correlate with tracer species

They are: (1) fast sensors on mobile platforms; (2) remote sensors; and (3) distributed sensor networks. These strategies are listed in Box 1, which includes some descriptors of their current or potential implementations. Of course, these strategies are not mutually exclusive, remote sensing instruments can be mounted on mobile platforms, with airborne and space-based passive (radiometers, spectrometers) and active (lidar, radar) remote sensing instruments playing a large role in our understanding of global-scale environmental issues such as stratospheric ozone destruction and climate change. Further discussion of these three approaches, their application to a variety of environmental problems, and the instrument challenges they pose can be found in the report of a recent workshop on environmental instrumentation sponsored by the National Science Foundation.[3] The environmental science community has made enormous advances in

---

[3]*Instrumentation for Environmental Science 2000—Report of a Workshop and Symposium,* National Science Foundation, Arlington, VA, 2000.

BOX 1
Strategies for Solving the Undersampling Problem

Mobile Platform with Fast Sensors (Process Studies)
   Platforms: manned-unmanned aircraft, balloons, trains, vans
   In situ or remote sensors with 0.01 to 10 second response
Remote Sensing (Trends and Distributions)
   Satellite or airborne passive or active optical instruments
Distributed Sensor Networks (Trends and Distributions)
   Large numbers of relatively inexpensive sensors linked by
   telecommunication (wireless, fiber optic) networks

---

the past two decades in designing, fielding, and utilizing rapid sensors on mobile platforms and remote sensing instruments. One point made in the NSF report is that anticipated advances in sensors based on micro- and nanotechnology—coupled with advanced information technology solutions to the problems of data collection, processing, dissemination, and display—may make the distributed sensor network measurement strategy much more powerful and affordable.

Over the past 15 years, the atmospheric science community has developed a series of mobile platforms with highly accurate and specific fast response instrumentation that have revolutionized atmospheric chemistry field measurements. These include high-altitude aircraft, such as NASA's ER-2 and WB-57, and lower-altitude aircraft like the NASA DC-8, the National Oceanic and Atmospheric Administration (NOAA) and Center for Interdisciplinary Remotely-Piloted Aircraft Studies (CIRPAS) (Naval Postgraduate School) Twin Otters, the National Center for Atmospheric Research (NCAR) C-130, and the DOE G1. In addition, mobile surface laboratories are now being used for a wide variety of urban and regional air quality and emission source characterization studies.[4] Typical configurations for the ER-2 and the mobile laboratory are shown in Figures 1 and 2.

Interestingly, the mobile (comparatively) fast-sensor strategy has recently been implemented for subsurface studies in the form of a truck-mounted geotechnical probe equipped with miniaturized laser-induced fluorescence or laser-induced breakdown spectroscopy sensors to detect vertical and horizontal dis-

---

[4]See for example, Lamb, B.; McManus, J. B.; Shorter, J. H.; Kolb, C. E.; Mosher, B.; Harriss, R. C.; Allwine, E.; Blaha, D.; Howard, T.; Guenther, A.; Lott, R. A.; Siverson, R.; Westberg, H.; Zimmerman, P. *Environ. Sci. Technol.* **1995,** 29, 1468-1479; Jiménez, J. L.; McManus, J. B.; Shorter, J. H.; Nelson, D. D.; Zahniser, M. S.; Koplow, M.; McRae, G.J.; Kolb, C. E. *Global Change Sci.* **2000,** *2,* 397-412.

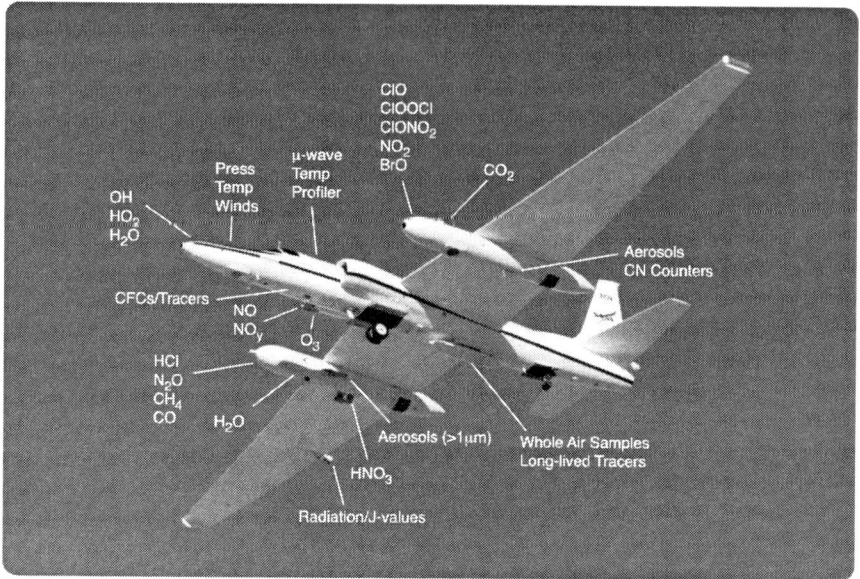

FIGURE 1 View of NASA WER-2 High Altitude Aircraft With Stratospheric Chemistry Instrument Package.

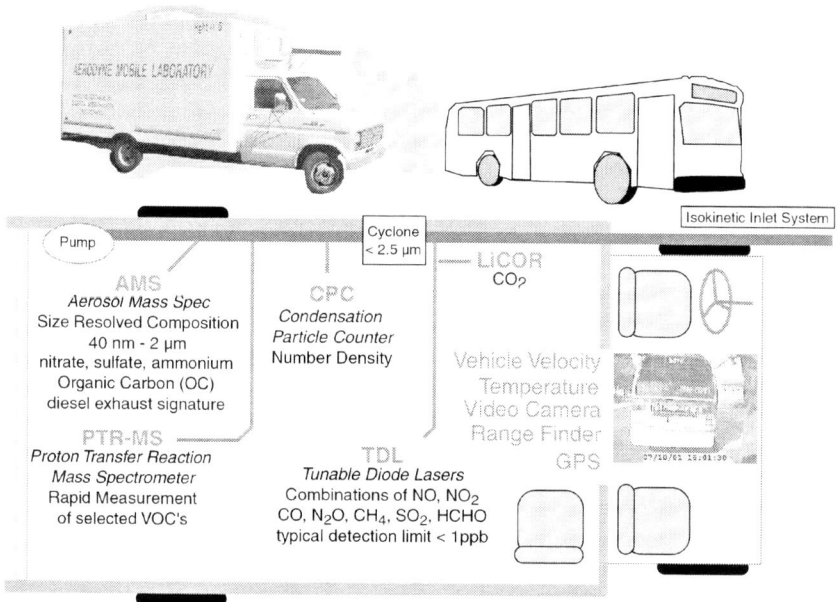

FIGURE 2 Schematic of aerodyne research mobile laboratory deployed for air quality studies in Mexico City.

BOX 2
Innovative Atmospheric Chemistry Field
Measurement Design

Goal
Exploit capabilities of advanced instruments and platforms to form effective and responsive field measurement systems tailored to the issues being addressed.

Field Measurement System Characteristics
- Simultaneous real-time measurement of many trace species
- Each measurement technique calibrated and quality assured
- Mobile, fast-response instruments replace multiple, slow, fixed-site monitors
- Advanced platforms cover spatial scales necessary to address problem
- Careful blending of static and mobile point and remote sensing measurements to assure intercompatibility and synergy
- Immediate model analysis of field data to assess measurement success and scope
- Ongoing measurement efforts

---

tributions of aromatic organic and heavy metal contaminants, respectively.[5] Box 2 lists the general goal and some of the desirable characteristics for exploratory and process study field measurement campaigns that are required to take full advantage of fast-sensor–mobile platform strategy. Box 3 illustrates some of the evolution in environmental analytical instrumentation (especially in atmospheric measurement technology) that has motivated the development of instrument suites such as those illustrated in Figures 1 and 2.

In order to produce the instruments and platforms that meet the specifications required by the measurement strategies listed in Box 1, advances in enabling technologies must be exploited. Table 2 lists some of the enabling technologies that are currently undergoing rapid development and presents examples of the improvements in environmental instrumentation and measurement platforms they have or may soon allow.

---

[5]See for example, Sinfeld, J. V.; Germaine, J, T.; Hemond, H.F. *J. Geotech Geoenviron.* **1999,** *125,* 1072-1077; Theriault, G. A.; Bodensteiner, S.; Liberman, H. *Field Anal. Chem. Technol.* **1998,** *2,* 117-125; *Instrumentation for Environmental Science 2000—Report of a Workshop and Symposium,* National Science Foundation, Arlington, VA, 2000.

BOX 3
Atmospheric Species Measurements Evolution

Measurement Science: Wet Chemistry → Electro- or Ion Optics
Sampling Frequency: Batch → Continuous
Analysis Rate: Off-Line → Real Time
Sample Preparation: Concentration-Conversion → Whole Air
Specificity: Species Classes → Species → Isotopomers

TABLE 2 Impacts of Enabling Technologies

Technology	Examples of Evolving Impacts
Structural materials	Lighter, more robust sensors and sensor platforms
Energy systems	Smaller, longer lasting off-grid power sources, enhanced platform propulsion systems
Electro-optics	Compact, efficient solid state lasers and detectors, compact long-path sampling cells, detector arrays
Semiconductor technology	Compact, robust electronics, throw-away sensors and data systems
Ion optics	Ion traps, smaller TOF and quadrupole mass filter spectrometers
Vacuum technology	Fieldable sensors based on electron, ion, molecular, particle beam methods
Information technology	Real-time data processing and display, multisensor integration, data fusion and assimilation
Control technology	Autonomous instrument-platform operation, real-time experimental design
Fluid dynamics	More efficient airborne platforms, better sampling systems
Biotechnology	Smaller, faster diagnostics for microbial and biologically active molecules
Nanotechnology	Smaller, lighter, cheaper everything

Figure 3 shows an airborne version of one recently developed fast response instrument made possible by recent advances in materials, vacuum technology, ion optics, fluid dynamics, information technology, and control technology.[6] This aerosol mass spectrometer allows the real-time measurement and display of the nonrefractory, size-resolved (~30 nm to ~1500 nm) ambient aerosol particle mass loadings. Figure 4 shows the ambient fine aerosol nonrefractory composition,

---

[6]Jayne, J. T.; Leard, D. C.; Zhang, X.; Davidovits, P.; Smith, K. A.; Kolb, C.E.; Worsnop, D. R *Aerosol Sci. Technol.* **2000,** *33,* 49-70.

Physical
  Power: *600 watts*
  Mass: *190 kg*
  Size: *~36" x 36" x 24"*
          (plus half rack of electronics)
Performance
  Measurement size range: 30 nm
    to 1 μm
  Sensitivity: *0.01 μg/m³ (60s sec.)*
  Sampling Rate: *100 cm³ min⁻¹*
  Maximum Data Rate: *100 Hz*

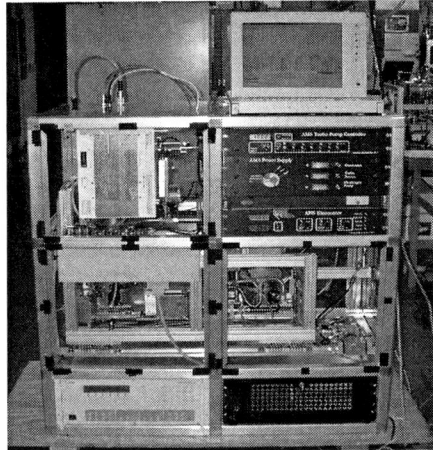

G1 Package
*Integral size and composition information for
non-refractory bulk constituents
on seconds time scale*

FIGURE 3 An Aerosol mass spectrometer allowing real-time size-resolved chemical speciation measurements of fine aerosols, free standing and packaged for airborne measurements.

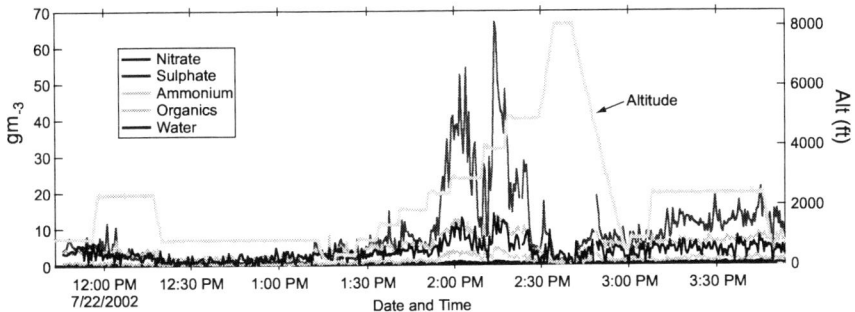

FIGURE 4 Aerosol mass spectrometer airborne measurements of fine particle compositional mass loading and size distributions showing two sulfate aerosol layers in air masses
from the Ohio River valley.

mass loading, and size distribution measured as a function of aircraft altitude of the coast of Massachusetts in July 2002. The two large sulfate peaks are sampling air with back trajectories that left the surface over the Ohio River valley two days earlier.

Summary

It is clear that innovative instruments and measurement strategies are required to cure the undersampled environment problem. They can be developed and deployed if ongoing advances in a wide range of enabling technologies are adapted and exploited. For many environmental challenges, measurement requirements favor real-time, mobile, autonomous instruments. The resulting quantum leaps in measurement capabilities will have the potential to revolutionize atmospheric and environmental chemistry.

ENVIRONMENTAL BIOINORGANIC CHEMISTRY

François M.M. Morel
Princeton University

The coevolution of life and geochemistry on the Earth has resulted in an extraordinarily tight coupling between the cycles of bioactive elements at the surface of our planet and the growth of microorganisms. This is evidently true of major plant nutrients such carbon, nitrogen and phosphorus. It is also true of many essential trace metals, such iron, manganese, copper, cobalt, nickel, and zinc, that serve as active centers in enzymes that catalyze the transformations of carbon and nitrogen in terrestrial and aquatic systems. The biological availability and use of these metals are modulated by intracellular and extracellular binding compounds produced by microorganisms. The chemistry of bioactive metals in the environment, including their coordination to binding agents and to appropriate centers in proteins, thus controls the efficiency of critical environmental processes that govern the global cycles of carbon and nitrogen. A detailed understanding of the *bioinorganic chemistry* of these metals in various environs and of their effects on geochemically important processes presents major challenges and opportunities to environmental chemists as illustrated in the three oceanographic examples below.

The increase in O_2 and concomitant decrease in CO_2 caused by the evolution of oxygenic autotrophs on the Earth, has resulted in an undersaturation of the main (and ancient) carboxylating enzyme, ribulose-1,5-bisphosphate carboxylase, responsible for the first step in carbon fixation—the dark reaction of photosynthesis. To palliate this difficulty, a number of species of marine phytoplankton have evolved carbon concentrating mechanisms that all involve some forms of the zinc enzyme carbonic anhydrase.[1] This enzyme catalyzes the hydration of CO_2 and the dehydration of HCO_3^-. In the surface oceans, which are singularly depleted in zinc, a number of phytoplankton species have evolved the ability to replace zinc with cobalt and cadmium in carbonic anhydrase.[2,3] In diatoms, which are arguably the most important primary producer in the modern oceans, it appears that the activity of the external carbonic anhydrase is enabled by the formation of a silica frustule that serves, in part, as a proton buffer at the surface of the cell.[4] Unraveling the conditions that allow for an efficient Zn-Co-Cd replacement in carbonic anhydrase (and similar enzymes) in marine autotrophs and of the use of silica as a local proton buffer will provide a molecular understanding of the links between the global cycles of these trace metals and of CO_2 and SiO_2.

The increase in O_2 concentration in the oceans since the evolution of oxygenic photosynthesis has also led to a massive decrease in the concentrations of metals such as iron and manganese, which form insoluble oxy-hydroxides. These metals, particularly iron, are important in the transfer of electrons in photosynthesis and respiration and as metals centers in a wide variety of redox enzymes. Not surprisingly, there is mounting evidence that modern marine microorganisms wage a fierce "iron war" against each other and that many have acquired the ability to replace iron with other metals in key compounds. For example, the production of marine siderophores allows some marine bacteria to sequester iron and limit its availability to organisms that do not possess the appropriate transport proteins (Figure 1).[5] As a countermeasure, some eukaryotic phytoplankton extract the iron from Fe(III)-siderophore complexes by reduction of Fe(III) to Fe(II) extracellularly.[6] We need to elucidate the chemical forms of iron in seawater and the mechanisms of iron acquisition by various microorganisns in order to assess the role of iron in limiting primary production in the oceans[7] and in controlling the assemblage of planktonic species.

Nitrogen is generally considered the most important limiting plant nutrient in the oceans. The processes that determine the total concentration of fixed nitrogen in surface seawater—particularly N_2 fixation and denitrification—thus exert a fundamental control on marine primary production and on the resulting seques-

[1]Badger, M. R.; Hanson, D.; Price, G. D. *Functional Plant Biology* **2002**, *29*, 161-173.

[2]Morel, F. M. M.; Reinfelder, J. R.; Roberts, S. B.; Chamberlain, C. P.; Lee, J. G.; and Yee, D. *Nature* **1994**, *369*, 740-742.

[3]Lane, T. W.; Morel, F. M. M. *Proc. Natl. Acad. Sci. USA* **2000**, *97*, 4627-4631.

[4]Milligan, A. J.; Morel, F. M. M. *Science* **2002**, *297*, 1848-1850.

[5]Butler, A. *Science* **1998**, *281*, 207-210.

[6]Maldonado, M. T.; Price, N. M. *Journal of Phycology* **2001**, *37*, 298-309.

[7]Coale, K. H.; Fitzwater, S. E.; Gordon, R. M.; Johnson, K. S.; Barber, R. T. *Nature* **1996**, *379*, 621-624.

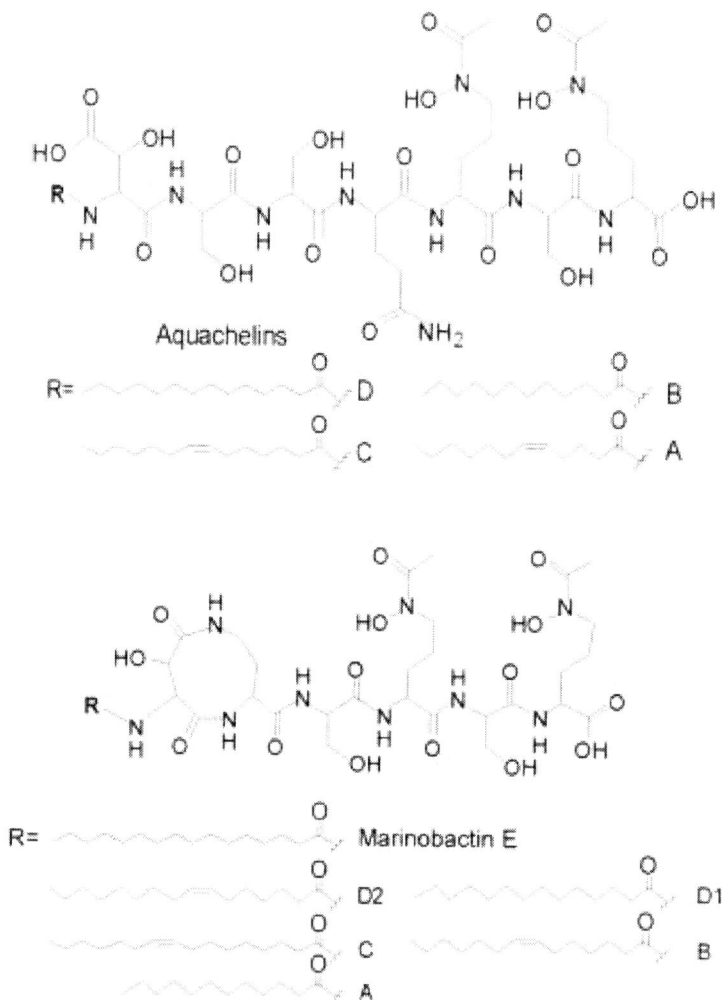

FIGURE 1 Two examples of ampiphilic siderophores produced by marine heterotrophic bacteria. Redrawn from Martinez et al. (2000).

tration of CO_2 in the deep oceans. All the steps in the nitrogen cycle are catalyzed by metalloenzymes (Figure 2). For example, all forms of nitrogenase (the enzyme responsible for N_2 fixation) contain a large number of iron atoms. Iron, molybdenum, and copper are also involved in the various enzymes that carry out denitrification. There is presently wide speculation that iron availability limits the overall

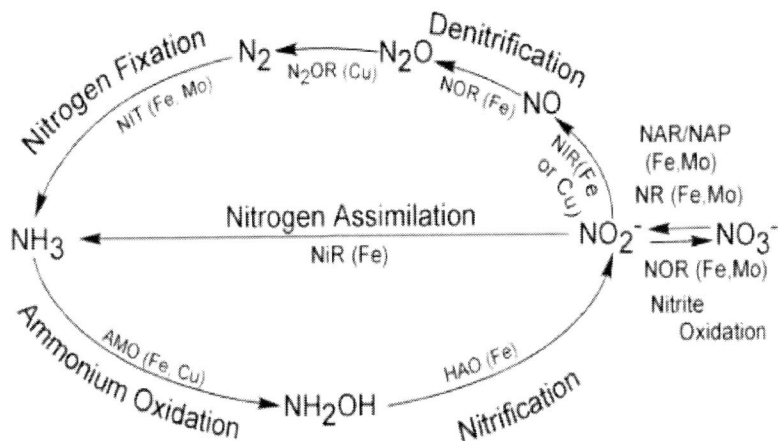

FIGURE 2 A diagram of the nitrogen cycle with catalyzing enzymes and metal requirements of each step. NOTE: AMO = ammonium mono-oxygenase; HAO = hydroxylamine oxidoreductase; NAR = membrane-bound respiratory nitrate reductase; NAP = periplasmic respiratory nitrate reductase; NR = assimilatory nitrate reductase; NIR = respiratory nitrite reductase; NiR = assimilatory nitrite reductase; NIT = nitrogenase; NOR = nitric oxide reductase; N_2OR = nitrous oxide reductase.

rate of nitrogen fixation in the oceans.[8,9] There is also some evidence that, in some suboxic waters, the concentration of available copper may be too low for the activity of the copper enzyme nitrous oxide reductase and result in N_2O accumulation and release to the atmosphere.[10] Understanding the nitrogen cycle of the oceans and the release of some important greenhouse gases such as N_2O to the atmosphere thus requires that we elucidate the acquisition of metals such as iron and copper and their biochemical utilization by various types of marine microbes.

As exemplified above, the major goals of environmental bioinorganic chemistry are to elucidate the structures, mechanisms, and interactions of important "natural" metalloenzymes and metal-binding compounds in the environment and to assess their effects on major biogeochemical cycles such as those of carbon and nitrogen. By providing an understanding of key chemical processes in the biogeochemical cycles of elements, such a molecular approach to the study of global processes should help unravel the interdependence of life and geochemistry on planet Earth and their coevolution through geological times.

[8]Falkowski, P. G. Nature 1997, 387, 272-275.

[9]Berman-Frank, I.; Cullen, J. T.; Shaked, Y.; Sherrell, R. M.; Falkowski, P. G. Limnology and Oceanography 2001, 46, 1249-1260.

[10]Granger, J.; Ward, B. Limnology and Oceanography 2002, 48, 313-318.

ENVIRONMENTALLY SOUND AGRICULTURAL CHEMISTRY: FROM PROCESS TECHNOLOGY TO BIOTECHNOLOGY

Michael K. Stern
Monsanto Company

Agricultural practices are undergoing a transformation driven partially by advances at the interface of chemistry and biotechnology. This paper outlines some of the new technologies that are playing an integral role in catalyzing that change. Topics of discussion include new chemical process technology and chemical catalysis that allows for more efficient production of herbicides as well as transgenic crops and their benefits to agriculture and the environment.

Glyphosate is the active ingredient in Roundup herbicide. The process Monsanto uses to manufacture glyphosate relies heavily on chemical catalysis (Figure 1). Improvements in our catalyst have driven profound environmental and economic benefits in the manufacturing of glyphosate. One of the main themes around environmental chemistry moving into the future will be a renewed focus on the development of novel homogeneous and heterogeneous catalysts. One of the beautiful things about catalysts is their ability to positively impact the economics of a process often without the requirement for major capital investments.

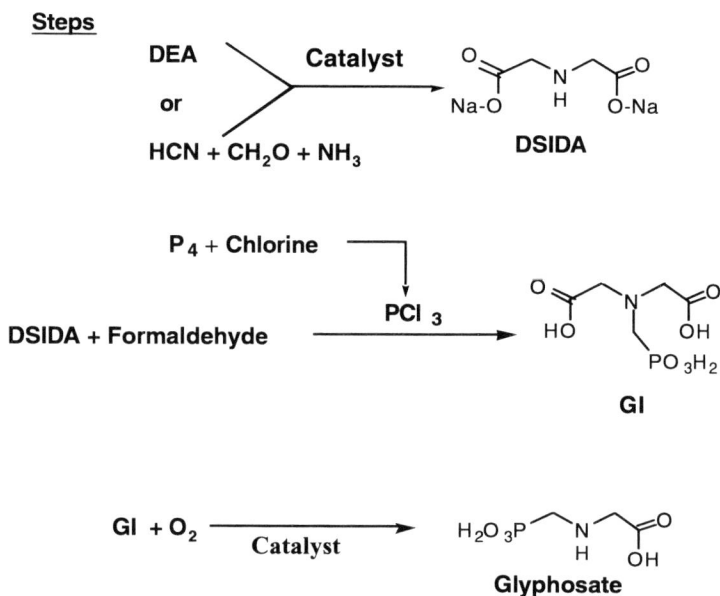

FIGURE 1 The current glyphosate process.

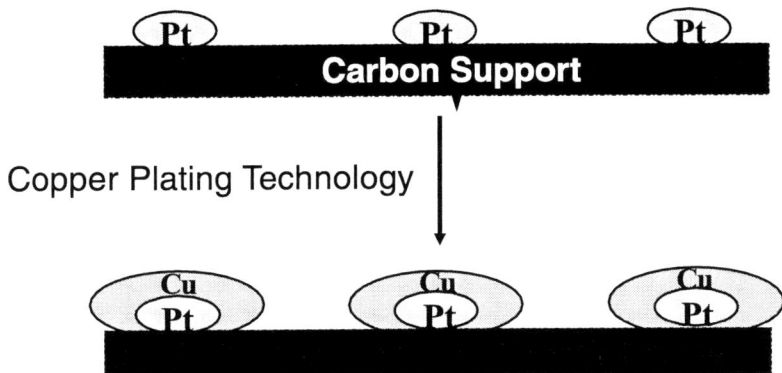

FIGURE 2 Second-generation catalyst.

A key intermediate in the manufacturing of glyphosate is disodium iminodiacetic acid (DSIDA). Originally an HCN-based route was the only process used to manufacture this product. Some of the advantages of this technology are (1) it's proven, (2) it uses readily available raw materials, and (3) the process typically gives good yields. However there are a lot of challenges associated with this chemistry. For instance, a considerable amount of waste is produced and HCN is a difficult raw material to handle due to its toxicity. Accordingly, Monsanto wanted to explore other technologies when we were faced with the need to expand our DSIDA capacity in the early 1990s.

We ultimately settled on a novel catalytic route for the production of DSIDA. The initial catalyst was a Raney copper composition that allowed us to convert diethanolamine, to DSIDA in a really interesting dehydrogenation reaction. This is an endothermic reaction that gives off hydrogen as a by-product. We were able to get the facility to work with this catalyst, but there were a lot of operational issues associated with this technology. Raney copper is a very malleable soft metal which resulted in catalyst stability issues. Ultimately we needed to go ahead and find something that was better.

Copper is essentially the only metal that catalyzes this reaction. The technical challenge was to find a way to stabilize copper under the reaction conditions. We spent a lot of time looking at whether you could put copper directly on carbon. It turns out that you really can't do that very well. Copper likes to move around, particularly under the reaction conditions. So a new catalyst technology was developed using platinum as an anchor that was then coated with copper. This resulted in a very stable catalyst (Figure 2). The new catalyst technology had significant environmental benefits. These included the use of less toxic raw materials and the elimination of nearly all of the waste produced by the older technology.

Let me switch gears to another catalytic reaction. This is the reaction where we take glyphosate intermediate and convert it to glyphosate using a carbon catalyst. This appears to be a very simple reaction, but there are several technical issues related to the production of by-products. The problem is that the by-products go ahead and react with your desired product to make other undesired by-products. What we needed was a catalyst that could do two reactions at the same time. The first reaction converts glyphosate intermediate to glyphosate and the other is to react away the undesirable by-products. We were successful in developing this type of catalyst which resulted in significant environmental and commercial benefits (Table 1).

The major benefits associated with this technology result from the more efficient use of water in the process. With the implementation of this new catalyst technology we have been able to reduce the amount of water flowing through our process by nearly 300 million gallons a year. This resulted in a concomitant reduction in flows to our biotreatment systems that reduce the amount of biosolids we send to landfill. Overall this was a very successful project for Monsanto.

I'd like to change focus from catalysis to another theme in green chemistry: that of atom efficiency. If you noticed in the glyphosate process, we're not completely atom efficient. This is due to the fact that putting on the phosphonomethyl group is challenging. The issue is that if you don't protect the primary amine, you end up doing two phosphonomethyl reactions that yield the undesirable product glyphosine. The current solution to this problem is to protect the primary amine with a carboxymethyl group, which ultimately gets removed in the last step of the process. The challenge was to find a more atom-efficient protecting group.

A team investigated this and developed a whole new technology based on novel platinum catalysts. This technology allows for the selective demethylation of N-methylglyphosate to produce glyphosate acid directly. This would save one

TABLE 1 Environmental Benefits of New Catalyst and Process Technology

	Annual Reductions Projected by 2002
Resources	
Steam (BTUs/yr)	880,000,000,000
Demineralized water (gal/yr)	380,000,000
Waste	
Flow to biosystems (gal/yr)	800,000,000
SARA 313 deep well injection (lb/yr)	52,000
Biosludge (lb/yr)	8,000,000
Land-filled solid waste (lb/yr)	1,380,000
SARA 313 air emissions (lb/yr)	17,600
Carbon dioxide production (lb/yr)	100,000,000

NOTE: SARA = Superfund Amendments and Reauthorization Act.

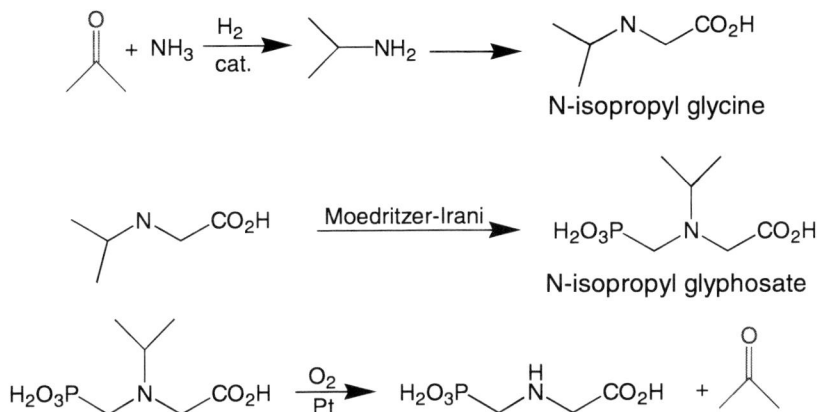

FIGURE 3 *N*-isopropyl glyphosate: an atom-efficient intermediate.

carbon atom when compared to the traditional route. However it was also discovered that you could use other protecting groups besides methyl. In fact it was possible to use an isopropyl group, which under the reaction conditions could be removed to generate acetone and glyphosate. The acetone can be recycled back into the process, resulting in an extremely atom-efficient system (Figure 3).

As mentioned above, glyphosate is the active ingredient in Roundup herbicide. Roundup plays an important part in the new wave of agricultural products derived from biotechnology. This new technology has many economic and environmental benefits. Expansion of the global acreage planted with Roundup Ready crops has resulted in a reduction of the use of pesticides by nearly 50 million pounds per year. This can affect groundwater positively by reducing agricultural chemical contamination in watersheds where a large percentage of Roundup Ready crops are planted. When crops such as cotton and corn are protected against insect pest through biotechnology, we also see a benefit to nontarget organisms. So in summary, biotechnology has delivered significant environmental benefits. Many of these benefits are consistent with the EPA's guidelines and focus.

In conclusion, the chemical industry is going to be even more dramatically transformed in the future, and key advances will be made at the interface of chemistry and biology. Discovery and development of new environmentally beneficial catalyst and process technologies will be critical for the chemical industry to thrive in the United States. The new technologies will need to be relevant and have a positive impact on the earnings and competitiveness of the chemical industry. Advances in catalysis will be a key driver in the development of cleaner and more efficient chemical processes. Finally, breakthrough discoveries are likely to be the products of interdisciplinary work teams.

STABLE ISOTOPES AND THE FUTURE OF ENVIRONMENTAL-CHEMICAL RESEARCH

Mark H. Thiemens
University of California, San Diego

In 1947, Harold Urey and, simultaneously, Bigeleisen and Mayer developed the formalism for determination of the position of equilibria in isotope exchange reactions. These papers, for the first time, calculated at high precision the position of isotope exchange equilibria as a function of temperature. In the same year, Nier reported the development of the double collector isotope ratio mass spectrometer, which allowed for measurements of isotope ratios at a precision sufficient to measure the modest isotope ratio changes associated with the temperature dependencies of exchange reactions. In this sense, 1947 represents the birth of stable isotope chemistry. Subsequently, an enormous range of applications has emerged utilizing isotope ratio measurements as a probe of natural processes that include studies of atmospheric chemical processes, paleoceanography and climate, stable isotope geochemistry, and planetary sciences.

Although the utilization of stable isotopes as a means to resolve environmental processes has had many applications for more than a half-century, there have been some limitations on the extent to which specific processes, both chemical and physical, might be resolved. These limits arise because with only a single isotope ratio, there is a certain lack of specificity associated with the measurements.

In 1983, Thiemens and Heisenreich reported a new isotope effect. This particular effect was unique in that the isotope ratios (e.g., of oxygen) alter on a basis other than mass. For example, in the case of ozone formation, the isotopomers

$$^{16}O^{16}O^{17}O$$
$$^{16}O^{16}O^{18}O$$

form at essentially equally rates that exceed those associated with

$$^{16}O^{16}O^{16}O.$$

These reactions proceed in part on the basis of isotopic symmetry, with the asymmetric species forming at a rate greater than the purely symmetric species. Marcus and colleagues at Cal Tech have developed a chemical formalism that accounts for a large proportion of the laboratory experiments. There remain some fundamental issues, however, with respect to a fully developed quantum-level theory.

While additional theoretical formalisms for the isotopic fractionation event are still needed, the mass-independent isotopic fractionation process has provided a new and definitive mechanism by which an extraordinary range of environmental processes may be resolved. The inclusion of the second isotope ratio measurement adds a sensitive probe of processes that may not be afforded by concentration or single isotope ratio measurements. As such, the use of mass-independent isotope compositions of environmental molecular species has provided a new probe to understand natural processes and to characterize anthropogenic impacts. There will be a wide range of new and significant applications in the future that will provide a powerful complement to other measurement techniques and modeling efforts. These include studies of the atmosphere, hydrosphere, and geosphere, as well as paleoenviroments and global environmental change. Such studies will be of critical importance in evaluating and predicting global environmental sustainability.

Present and Future Applications

There exist numerous recent review articles on the subject of mass-independent isotope effects and details are available in these articles.[1,2,3] A key point of the culmination of observations is that the most important chemical issues in the environment today, and as expected in the future, may be studied utilizing mass-independent isotopic compositions. In the context of this report, this represents a future chemical frontier area of chemical-environmental research.

There exist a number of gases that possess mass-independent isotopic compositions, and with future measurements, the frontiers of environmental chemistry may be extended. The following are specific, though not inclusive, examples.

Stratospheric and Mesospheric CO_2

It was first observed by Thiemens et al.[4] that stratospheric CO_2 possesses a large and variable mass-independent isotopic composition. This composition was suggested as deriving from isotopic exchange with $O(^1D)$, the product of ozone photolysis.[5,6] As later confirmed by rocket-borne collection of stratospheric and mesospheric air, this unique isotopic signature provides an ideal tracer of odd oxygen chemistry of the Earth's upper atmosphere, one of the most important upper atmospheric processes. There are, however, several features that require further measurement (laboratory and atmospheric) and theoretical considerations.

[1]Thiemens, M. H. *Science* **1999**, *283*, 341.

[2]Weston, R. E. *Chem. Rev.* **1999**, *99*, 2115.

[3]Thiemens, M. H.; Savarino, J.; Farquhar, J.; Bao, H. *Acct. Chem. Res.* **2001**, *34*, 645.

[4]Thiemens, M. H.; Jackson T. L.; Mauersberger, K.; Schuler, B.; Morton, J. *Geophys. Res. Lett.* **1991**, *18*, 669.

[5]Yung, Y. L.; DeMore; W. B.; Pinto; J. P. *Geophys. Res. Lett.* **1991**, *18*, 13.

[6]Yung, Y. L.; Lee, A. Y. T.; Iriow, W. B.; Demaro, W. B.; Chen, J. *J. Geophys. Res.* **1997**, *102*, 10857.

This represents a future goal in need of pursuit because both climate and chemical budgeting considerations are impacted. As discussed in an earlier report,[7] upper atmospheric CO_2 possesses a mass-independent isotopic composition, while tropospheric CO_2 is strictly mass dependent (as a result of equilibrium isotopic exchange with water). This renders CO_2 an ideal tracer of stratosphere and troposphere mixing. Quantification of this process is also of significance for understanding global chemical budgets and lifetimes of several greenhouse species. Given that significant uncertainties remain, future measurements will be crucial in quantification of stratospheric and tropospheric mixing and in development of remediation policies associated with global climate change.

Greenhouse Gas Characterization

As discussed in the review article,[8] several greenhouse gases have been observed to possess mass-independent isotopic compositions. These include O_3, CO, and N_2O. In each instance, this measurement has provided significant insight unattainable by other measurement techniques. The case of atmospheric N_2O is particularly interesting. Nitrous oxide is a greenhouse gas with a warming capacity nearly 200 times that of CO_2 on a per-molecule basis, and serves as a major sink for stratospheric ozone via photochemical destruction. In spite of decades of research, the N_2O budget is still inadequately understood. Most recently, isotopic measurements of a new variety have proven to be particularly valuable. These observations utilize high-precision measurements of the isotopomeric fragments of NO in a mass spectrometer.[9,10,11,12] From such measurements, the internal N_2O isotopomeric distribution may be determined. It is now recognized that the isotopomeric distributions of

$$^{15}N^{14}N^{16}O$$
$$^{15}N^{14}N^{18}O$$
$$^{14}N^{15}N^{16}O$$
$$^{14}N^{15}N^{18}O$$

[7]Thiemens, M. H.; Jackson, T.; Zipf, E.C.; Erdman, P.W.; Van Egmond, C. *Science* **1995**, *270,* 969.

[8]Thiemens, M. H. *Science* **1999**, *283,* 341.

[9]Toyoda, S.; Yoshida, N. *Anal. Chem.,* **1999**, *71,* 4711-4718.

[10]Brenninkmeijer, C. A. M.; Rockmann, T. *Rapid Commun. Mass Spec.* **1999**, *13,* 2028.

[11]Rockmann, T.; Kaiser, J.; Crowley, J. W.; Brenninkmeijer, C. A. M.; Crutzen, P. *Geophys. Res. Lett.* **2001**, *28,* 503.

[12]Toyoda, S.; Yoshida, N. C.; Urabe, T.; Aoki, S.; Nakazawa, T.; Sugawara, S.; Honde, H. *J. Geophys. Res.* **2001**, *106,* 7515.

are highly characteristic of specific processes, such as photolysis or defining individual point sources (biologic, abiologic).

In the future, this particular variety of measurements will provide a new level of detail in understanding the global atmospheric cycles of N_2O as well as the other gases. There is a clear need for expansion of such measurements in a variety of global environments and to obtain fundamental physical-chemical information simultaneously. With such concomitant developments, new details of N_2O atmospheric processes may be obtained.

Atmospheric Aerosol Sulfate and Nitrate

Atmospheric sulfate aerosols are known to exert a significant influence on Earth's surficial processes. They mediate climate, both as cloud condensation nuclei and as light-scattering agents. Sulfate is also a pernicious respirable molecule with well-known consequences for human health. It is estimated that there are more than 60,000 deaths a year from cardiovascular disease associated with aerosol inhalation. Recent studies have demonstrated a link to cellular damage. Additionally, following wet and dry deposition, sulfate destroys biota, alters biodiversity, and causes pervasive structural damage. The segregation of sulfur between gas- and aqueous-phase oxidative processes has implications for the degree of indirect effect of sulfate aerosols on climate due to its dependence on aerosol number densities. Sulfate derived from gas-phase oxidation results in new particle formation. However, for aqueous-phase oxidation, the sulfate is generated on a previously existing particle and does not contribute to the total aerosol number.

It has recently been shown that atmospheric sulfate possesses a significant and variable oxygen mass-independent isotopic composition. From a series of laboratory and atmospheric measurements it has been demonstrated that these measurements provide a highly sensitive mechanism by which the relative proportions of hetero- and homogeneous oxidative pathways may be quantified. This represents a significant observational advancement, and future measurements on a global scale will dramatically enhance understanding of this important atmospheric species. It has also been shown that sulfate oxygen isotopic measurements of aerosols collected during the Indian Ocean Experiment (INDOEX) revealed that the Intertropical Convergence Zone (ITCZ) is a source of new aerosol particles, which has significant consequences. First, this is a previously unrecognized process unaccounted for in any global climate model. Secondly, this is a source of large (micron-sized) particles and the mechanism associated with their formation is unknown. Gas-to-particle conversion processes produce submicron- rather than micron-sized particles. The likely mechanism for this process may be surface catalysis, possibly on carbonaceous particles. Should this be confirmed, there would be significant consequences. Climate models assume that sulfate particles are white and reflective of visible light. If sulfur is catalytically oxidized on carbon surfaces, the sign of sulfate radiative forcing may be the reverse of what has been assumed, in which case there could be significant error in climate models. It is therefore of considerable importance that the nature and magnitude of this process be resolved. Fully understanding the reaction pathways can be accomplished only by an intensive combination of isotopic measurements, climate models, and laboratory studies of the relevant surface catalytic reactions. Such a program typifies the future needs of environmental chemical research.

As is the case of sulfate, nitrate aerosols also possess large, mass-independent isotopic compositions. Nitrate concentrations may double in the next half-century with severe environmental consequences, including alteration of biodiversity, enhanced algal blooms, and loss of agricultural productivity. As in the case of sulfate, the large mass-independent isotopic signature has afforded new insights into a major global cycle—the nitrogen cycle. There remains a large range of studies to be pursued, which will develop new understanding of the chemistry of the Earth's environment. Once more, this may be accomplished by combined laboratory physical-chemical measurements, field observations, and modeling efforts. The ultimate consequence is a significant advancement in understanding global interactions of the chemical environment.

Summary

The utilization of mass-independent isotopic measurements of a wide variety of atmospheric, hydrospheric, and geologic species has advanced understanding of a wide range of environmental processes. The future development of the utilization and understanding of this new technique clearly will have numerous applications that should, and will, be advanced. Issues in climate change, health, agriculture, biodiversity, and water quality all may be addressed. Simultaneous with the acquisition of new environmental insight will be enhanced understanding of fundamental chemical physics.

E

Biographies of Workshop Speakers

James G. Anderson is Philip S. Weld Professor of Atmospheric Chemistry at Harvard University. He received his B.S. in physics from the University of Washington and his Ph.D. in physics-astrogeophysics from the University of Colorado. His research addresses three domains within physical chemistry: (1) chemical reactivity viewed from the microscopic perspective of electron structure, molecular orbitals, and reactivities of radical-radical and radical-molecule systems; (2) chemical catalysis sustained by free-radical chain reactions that dictate the macroscopic rate of chemical transformation in the Earth's stratosphere and troposphere; and (3) mechanistic links between chemistry, radiation, and dynamics in the atmosphere that control climate. Studies are carried out both in the laboratory, where elementary processes can be isolated, and within natural systems, in which reaction networks and transport patterns are dissected by establishing cause and effect using simultaneous, in situ detection of free radicals, reactive intermediates, and long-lived tracers. Professor Anderson is a member of the National Academy of Sciences.

Thomas W. Asmus is a senior research executive at DaimlerChrysler Corporation. He is the corporate adviser on powertrain technologies and key corporate technical representative in industry-government joint technical activities. He represents most of the research activities at the Chrysler Group. He has been involved in many facets of engine R&D activities including energy management systems, alternative engines, emissions control, and fuel economy improvement for nearly 30 years. Dr. Asmus is a Society of Automotive Engineers (SAE) fellow and received the 1999 American Society of Mechanical Engineers (ASME) Soichiro Honda Lecture Award, which was established by ASME to recognize individual achievement or outstanding contributions in the field of personal transportation. Dr. Asmus is a member of the National Academy of Engineering.

Ruben G. Carbonell is Frank Hawkins Kenan Distinguished Professor of Chemical Engineering at North Carolina State University. He is currently director of the William R. Kenan, Jr., Institute for Engineering, Technology, and Science at North Carolina State University, codirector of the National Science Foundation (NSF) Science and Technology Center for Environmentally Responsible Solvents and Processes, and director of the Kenan Center for the Utilization of Carbon Dioxide in Manufacturing. His main areas of research include the application of transport phenomena and colloid and surface science principles to the development of coating, extraction and reaction processes based on carbon dioxide as a solvent, and affinity chromatographic separation processes for biological molecules using ligands derived from combinatorial libraries. He earned his B.S. in chemical engineering from Manhattan College and Ph.D. in chemical engineering from Princeton University. He won the North Carolina State University Alumni Association Outstanding Research Award in 1989; the R.J. Reynolds Award for Excellence in Teaching, Research, and Extension from the North Carolina State College of Engineering in 1990; and the Alcoa Outstanding Research Award from the College of Engineering in 2001.

Uma Chowdhry is vice president for Central Research and Development (CR&D) at the DuPont Company. She joined DuPont in 1977 as a research scientist in CR&D and spent the first 11 years of her career there in various research and management roles. She subsequently served as laboratory director and business manager in the Electronics Department before moving to the Chemicals Sector as laboratory director of Jackson Lab. From 1993 to 1995, she served as R&D director for Specialty Chemicals and was appointed business director for the DuPont Terathane business in 1995. In 1997, Dr. Chowdhry moved back to Specialty Chemicals as business planning and R&D director. In 1999, she was appointed director of DuPont Engineering's Technology Division. Born and raised in Mumbai, India, she came to the United States in 1968 with a B.S. in physics from the Indian Institute of Science, Mumbai University; she received an M.S. from Caltech in engineering science and a Ph.D. in materials science from the Massachusetts Institute of Technology (MIT). Dr. Chowdhry is a fellow of the American Ceramic Society and a member of the National Academy of Engineering.

Barry Dellinger is the Patrick F. Taylor Chair of the Environmental Impact of Treatment of Hazardous Wastes and professor of chemistry at Louisiana State University (LSU). He is the director of the LSU Intercollege Environmental Cooperative and the acting director of the Biodynamics Institute. He is a member of the U.S. Environmental Protection Agency (EPA) Science Advisory Board Environmental Engineering Committee. From 1981 to 1998, he was group leader of environmental sciences and engineering at the University of Dayton where he also held a joint faculty appointment. From 1978-1981 he was a senior project scientist at Northrop Services Inc. He holds a Ph.D. in physical chemistry from Florida State University and a B.S. in chemistry from the University of North Carolina at Chapel Hill. His research interests include origin and control of toxic combustion by-products, thermal treatment of hazardous wastes, pathways of formation of dioxins, gas-phase and surface-catalyzed elementary reaction kinetics, and sources and health impacts of environmentally persistent free radicals. He is a recipient of the Charles A. Lindberg Certificate of Merit, the Engineering and Science Foundation Award for Outstanding Professional Achievement, the Wohleben-Hochwald Researcher of the Year Award, the Ohio General Assembly Award for Research Excellence, and corecipient of numerous EPA Science to Achieve Results (STAR) research awards.

David A. Dixon is a Battelle fellow in the Fundamental Science Directorate at the Pacific Northwest National Laboratory (PNNL), where he previously served as associate director for theory, modeling, and simulation at the William R. Wiley Environmental Molecular Sciences Laboratory. His main research interest is the use of numerical simulation to solve complex chemical problems with a primary focus on the quantitative prediction of molecular behavior. He uses numerical simulation methods to obtain quantitative results for molecular systems of interest to experimental chemists and engineers with a specific focus on the design of new materials and production processes. Before moving to PNNL, he was research fellow and research leader in computational chemistry at DuPont Central Research and Development (1983-1995) and a member of the Chemistry Department at the University of Minnesota, Minneapolis (1977-1983). He earned his B.S. in chemistry from the California Institute of Technology and his Ph.D. in physical chemistry from Harvard University, where he served as a junior fellow of the Society of Fellows, Harvard University. He is a fellow of the American Association for the Advancement of Science, and a fellow of the American Physical Society. He is a recipient of the 1989 Leo Hendrik Baekeland Award presented by the American Chemical Society, the Federal Laboratory Consortium Technology Transfer Award (2000), and the 2003 American Chemical Society Award for Creative Work in Fluorine Chemistry.

William H. Farland is the acting deputy assistant administrator for science in the U.S. Environmental Protection Agency's Office of Research and Development. A member of the government's Senior Executive Service, Dr. Farland's permanent position is director of the EPA's National Center for Environmental Assessment, which has major responsibility for the conduct of chemical specific risk assessments in support of EPA regulatory programs, the development of agency-wide guidance on risk assessment, and the conduct of research to improve risk assessment. Dr. Farland began his EPA career in 1979 as a health scientist in the EPA's Office of Toxic Substances, while he continued his research endeavors at the George Washington University. His career has been characterized by a commitment to the development of national and international approaches to testing and assessment of the fate and effects of environmental agents.

Dr. Farland has received a number of awards and honors, including an EPA Silver Medal and several Bronze Medals, for this work. Dr. Farland holds a Ph.D. from the University of California, Los Angeles (UCLA) in cell biology and biochemistry, an M.A. in zoology from the same institution, and a B.S. from Loyola University, Los Angeles. He was awarded an Individual National Research Service Award from the National Cancer Institute to pursue postdoctoral training in DNA damage and repair at the University of California, Irvine, and at Brookhaven National Laboratory.

Janet G. Hering is a professor of environmental science and engineering at Caltech, where she has been a member of the faculty since 1996. Prior to that, she was an assistant and later associate professor of civil and environmental engineering at UCLA. She has an A.B. in chemistry from Cornell University, an A.M. in chemistry from Harvard University, and a Ph.D. in oceanography from the Massachusetts Institute of Technology-Woods Hole Oceanographic Institution Joint Program. She has published more than 30 papers in refereed scientific journals and is the coauthor of the book *Principles and Applications of Aquatic Chemistry.* She is a past recipient of the National Science Foundation's Young Investigator Award and Presidential Faculty Fellows Award and is a member of the editorial advisory board for the journal *Environmental Science & Technology.* Professor Hering's research interests include the biogeochemical cycling of trace elements in natural waters and water treatment technologies for the removal of inorganic contaminants from potable water. Her research includes both laboratory and field experimental studies and has been funded by the National Science Foundation, the U.S. Environmental Protection Agency, the American Water Works Association Research Foundation, California Sea Grant, the Petroleum Research Foundation, the University of California Water Resources Center, the Chevron Research and Technology Company, and the Metropolitan Water District of Southern California.

Charles E. Kolb is president and chief executive officer of Aerodyne Research, Inc. Since joining Aerodyne as a senior research scientist in 1971, his personal areas of research have included atmospheric and environmental chemistry, combustion chemistry, chemical lasers, materials chemistry, and the chemical physics of rocket and aircraft exhaust plumes. He is the author or coauthor of more than 150 archival publications in these fields. In the area of atmospheric and environmental chemistry, Dr. Kolb initiated Aerodyne's programs for the identification and quantification of sources and sinks of trace atmospheric gases and aerosols involved in regional and global pollution problems, as well as the development of spectral sensing techniques to quantify soil pollutants. He has also developed models of aircraft and rocket exhaust plume-wake chemical kinetics, condensation physics, and dispersion processes critical to the systematic assessment of the impact of aerospace systems on the chemical structure of the upper troposphere and stratosphere. Dr. Kolb received the 1997 Award for Creative Advances in Environmental Science and Technology from the American Chemical Society. He has been elected a fellow of the American Physical Society, the Optical Society of America, the American Geophysical Union, and the American Association for the Advancement of Science and has served as the atmospheric sciences editor of the journal *Geophysics Research Letters* (1995-1999).

François M.M. Morel is Albert G. Blanke Professor of Geosciences at Princeton University, where he is also director of the Princeton Environmental Institute and director of the Center in Environmental BioInorganic Chemistry. He received the Licence-ès-Sciences in Mathématiques Appliquées at the Université de Grenoble, the Diplome d'Ingénieur from Université de Grenoble, and a Ph.D. in engineering sciences from the California Institute of Technology, Pasadena. The main focus of his research is the interactions between aquatic microorganisms and their chemical milieu, with the goal of understanding the biogeochemical cycles of elements, particularly trace elements, and their ecological consequences at the local or global scale. The relationship between aquatic microorganisms and trace metals, some of which are necessary for growth and some of which are toxic, is important for the ecology of plankton, the environmental consequences of metal pollution, and the possible control of primary production. At present, he is particularly interested in testing the so-called *iron and zinc hypotheses,* which, if true, could have important consequences for the efficiency of CO_2 sequestration by the oceans and, hence, for the global carbon cycle.

Michael K. Stern is senior science fellow and director of technology, agricultural chemistry, at Monsanto Company. He joined Monsanto in of 1989 as a senior research chemist in the Metal Mediated Chemistry Group of Monsanto Corporate Research and was appointed science fellow in 1993 and senior science fellow in 1999. During his tenure at Monsanto, Dr. Stern has been actively involved in a variety of research programs relating to the development of novel chemical processes and pharmaceuticals. He was the recipient of the Monsanto Thomas and Hochwalt Science and Technology Award in 1993 for his role in the discovery of the aromatic substitution chemistry used in the development of Monsanto's new process to manufacture 4-aminodiphenylamine (4-ADPA) and received the Presidential Green Chemistry Challenge Award for this process technology in 1998. He graduated from Denison University with a B.S. in chemistry. He received an M.S. in chemistry from the University of Michigan and a Ph.D. in chemistry from Princeton University. He was a postdoctoral associate in the Department of Chemistry at MIT before joining Monsanto. Dr. Stern is the author of 24 scientific publications and the inventor on 24 U.S. Patents.

Mark H. Thiemens is dean of the Division of Physical Sciences and professor in the Department of Chemistry and Biochemistry at the University of California, San Diego (UCSD). He received his B.S. from the University of Miami and his Ph.D. from Florida State University. He joined the faculty at UCSD after serving as a postdoctoral fellow at the Enrico Fermi Institute at the University of Chicago. He has received an Alexander von Humboldt Fellowship, the E.O. Lawrence Award from the U.S. Department of Energy in 1999, and the Chancellors Associates Award for Excellence in Research from UCSD in 2001. He was named Chancellors Associates chair in 1999 and is a fellow of the American Academy of Arts and Sciences. His research is focused on measurement of stable isotope variations at ultrahigh precision to develop experimental programs in widely varying research fields that include atmospheric chemistry (aerosol and greenhouse gas studies), the evolution of life and the atmosphere, the physical chemistry of gas-phase photochemical reactions, early solar system history and Mars atmospherics (meteorite measurements), and gas-solid conversion mechanisms.

F

Workshop Participants

CHALLENGES FOR THE CHEMICAL SCIENCES IN THE 21ST CENTURY: WORKSHOP ON THE ENVIRONMENT

November 16-18, 2002

Hamid Abbasi, Gas Technology Institute
Heather C. Allen, Ohio State University
James G. Anderson, Harvard University
Ken B. Anderson, Argonne National Laboratory
Thomas W. Asmus, DaimlerChrysler Corporation
Roger Atkinson, University of California, Riverside
Betsey L. Ballash, University of California, Davis
Mark A. Barteau, University of Delaware
Philip H. Brodsky, Pharmacia (Retired)
Gordon E. Brown, Jr., Stanford University
William H. Brune, The Pennsylvania State University
Steven H. Cadle, General Motors
Ruben G. Carbonell, North Carolina State University
A. Welford Castleman, Jr., The Pennsylvania State University
Marge Cavanaugh, National Science Foundation
Uma Chowdhry, The DuPont Company
Helena Chum, National Renewable Energy Laboratory
Yoram Cohen, University of California, Los Angeles
Dady Dadyburjor, West Virginia University
Liese Dallbauman, Gas Technology Institute
Barry Dellinger, Louisiana State University

David A. Dixon, Pacific Northwest National Laboratory
Sandra K. Dudley, CH2M Hill
William H. Farland, U.S. Environmental Protection Agency
Howard J. Feldman, American Petroleum Institute
Allan M. Ford, Gulf Breeze, Florida
Joseph S. Francisco, Purdue University
Jean H. Futrell, Pacific Northwest National Laboratory
Raymond J. Garant, American Chemical Society
Richard A. Gross, Polytechnic University
Patrick G. Hatcher, The Ohio State University
George R. Helz, University of Maryland, College Park
John C. Hemminger, University of California, Irvine
Janet G. Hering, California Institute of Technology
Robert E. Huie, National Institute of Standards and Technology
William M. Jackson, University of California, Davis
Robert G. Keesee, University at Albany, SUNY
Charles Kolb, Aerodyne Research, Inc.
Stephen G. Maroldo, Rohm and Haas Company
Stephen W. McElvany, Office of Naval Research
Kristopher McNeill, University of Minnesota
Jay C. Means, Western Michigan University
Mario J. Molina, Massachusetts Institute of Technology
François M. M. Morel, Princeton University
Karl T. Mueller, The Pennsylvania State University
Mary Neu, Los Alamos National Laboratory
Parry M. Norling, RAND
Mitchio Okumura, California Institute of Technology
Charles Peden, Pacific Northwest National Laboratory
Jeffery P. Perl, Chicago Chem Consultants Corporation
Michael J. Prather, University of California, Irvine
Douglas Ray, Pacific Northwest National Laboratory
Sharon M. Robinson, Oak Ridge National Laboratory
J.W. Rogers, Jr., Pacific Northwest National Laboratory
Michael Romanos, University of Cincinnati
F. Sherwood Rowland, University of California, Irvine
Eric S. Saltzman, University of California, Irvine
John H. Seinfeld, California Institute of Technology
Brent Shanks, Iowa State University
Jeffrey J. Siirola, Eastman Chemical Company
Subhas K. Sikdar, U.S. Environmental Protection Agency
Christine S. Sloane, General Motors Corporation
Lynda Soderholm, Argonne National Laboratory
Michael K. Stern, Monsanto Company

Mark H. Thiemens, University of California, San Diego
Matthew Tirrell, University of California, Santa Barbara
Sam Traina, University of California, Merced
Frank P. Tully, U.S. Department of Energy
Richard P. Turco, University of California, Los Angeles
Jay R. Turner, Washington University
Jeanette M. Van Emon, U.S. Environnmental Protection Agency
Israel E. Wachs, Lehigh University
John C. Westall, Oregon State University
Elizabeth K. Wilson, Chemical & Engineering News

G

Reports from Breakout Session Groups

A key component of the Workshop on the Environment was the set of four breakout sessions that enabled individual input by workshop participants on the four themes of the workshop: discovery, interfaces, challenges, and infrastructure. Each breakout session was guided by a facilitator and by the expertise of the individuals as well as the content of the plenary sessions. Participants were assigned to one of four breakout groups on a random basis, although individuals from the same institution were assigned to different groups. Each breakout group (color-coded red, yellow, green, and blue) was asked to address the same set of questions and provide answers to the questions, including prioritization of the voting to determine which topics the group concluded were most important. After every breakout session, each group reported the results of its discussion in plenary session.

The committee has attempted in this report to integrate the information gathered in the breakout sessions and to use it as the basis for the findings contained herein. When the breakout groups reported votes for prioritizing their conclusions, the votes are shown parenthetically in this section.

DISCOVERY

What major discoveries or advances related to the environment have been made in the chemical sciences during the last several decades?

TABLE G-1 Organization of Breakout Sessions

Breakout Session	Group	Facilitator	Session Chair	Rapporteur
Discovery	Red	D. Raber	C. Sloane	John Westall
	Green	M. Barteau, P. Norling	M. Barteau, P. Norling	Mitchio Okumura
	Blue	P. Brodsky, W. Castleman	P. Brodsky, W. Castleman	Israel Wachs
	Black	J. Jackiw	J. Futrell	Kristopher McNeill
Interfaces	Red	J. Jackiw	J. Futrell	Kenneth Anderson
	Green	D. Raber	C. Sloane	Karl Mueller
	Blue	M. Barteau, P. Norling	M. Barteau, P. Norling	Heather Allen
	Black	P. Brodsky, W. Castleman	P. Brodsky, W. Castleman	Brent Shanks
Challenges	Red	P. Brodsky, W. Castleman	P. Brodsky, W. Castleman	Mary Neu
	Green	J. Jackiw	J. Futrell	William Brune
	Blue	D. Raber	C. Sloane	Jay Means
	Black	M. Barteau, P. Norling	M. Barteau, P. Norling	Douglas Ray
Infrastructure	Red	M. Barteau, P. Norling	M. Barteau, P. Norling	Robert Huie
	Green	P. Brodsky, W. Castleman	P. Brodsky, W. Castleman	Eric Saltzman
	Blue	J. Jackiw	J. Futrell	Gordon Brown
	Black	D. Raber	C. Sloane	Ray Garant

Red Group Report

Enabling Capabilities

Analytical (examples include single particles in the atmosphere, electron-capture detector, use of synchrotrons for aqueous systems, increased time resolution for atmospheric measurements)

Modeling (use of correlated chemical measurements to deduce environmental processes; modeling, from molecular to global; structure-activity relationships for activity and fate

Technical Solutions

Emission control of lean combustion; catalytic converter; substitutes—heavy metals, CFCs, water-based paints

Discoveries

Ozone hole; recognition of importance of speciation vs. total concentration; understanding of reactions on surfaces and microporous regions

Green Group Report

Tropospheric Chemistry

Chemistry of air pollution, (e.g., photochemical origin of smog; acid rain; discovery of the relevance of biogenic emissions; aerosol chemistry, formation, and microphysics)

Stratospheric Chemistry

The ozone hole—understanding of ozone chemistry in the stratosphere; heterogeneous chemistry in the atmosphere

Models, Databases, Techniques

Recognition of systems approach; generation of databases of kinetics and energetics; advances in computation; fundamental understanding of chemistry (free radicals); instrumentation (remote sensing, space based, advances in sensor technology)

Synthesis: Chemistry to Produce Systems with Minimal Environmental Impact

CFC replacements; PCB replacements; degradable pesticides, polymers; cleaner fuels; designer compounds

Process Advances

Implementing green chemistry; benign solvents; catalytic converters; engine efficiency via fundamental understanding; energy; energy storage and efficiency, photovoltaics; atom economy

Soil and Water Chemistry

Advances in interfacial chemistry, biogeochemical cycles; remediation technology; trace metals; radiation chemistry, radionuclides; hydrophobic compounds in the environment

Negative Advances

> MTBE
> Impact of catalytic self-propagating entities

Blue Group Report

Prioritized List

> Fundamental chemistry
> Analytical instrumentation
> Surface chemistry
> Genetics
> Understanding homogeneous atmospheric chemistry

Black Group Report

Tools

Industrial ecology; life-cycle analysis; detection, monitoring, measurement science (4); development of noninvasive spectroscopic technology; synchrotron-based methods, availability and development; application of GC-MS; computation and modeling (4); satellite technology (profiles, surface temperature, etc.) (3)

Technologies

Use of biomass in chemical reactions (instead of petroleum); advances in membrane science; metabolic engineering of microbes and plants for remediation; drinking water disinfection and ozonation technologies; three-way catalysts (1); emission control technology (3); use of supercritical CO_2 to replace solvents (1)

Risk Assessment

Development of structure-activity relationships; ability to predict risk and risk-based corrective action (2); identification of endocrine disrupters and elucidation of risk mechanism

Air Problems

Recognition of particulate matter (4); heterogeneous atmospheric chemistry (1); photochemical modeling ; understanding mechanisms of photochemical smog formation (2); rise and fall of CFCs; global warming; ozone depletion (3)

Water Problems

Groundwater remediation; bioaccumulation and persistent organic pollutants (2); detection of pesticides (1); human and eco-health effects of arsenic and mercury

Soil Problems

Discovery of dioxins
Discovery of PAHs

INTERFACES

What are the major environment-related discoveries and challenges at the interfaces between chemistry–chemical engineering and other disciplines, including biology, information science, materials science, and physics?

Red Group Report

Physics

Scaling of physical and chemical properties with size (10); instrumentation and sensor development (10); application of quantum mechanics to understand chemical processes (1); understanding transport coupling with chemistry (2)

Materials

Development of solvents; materials development (3); new catalysts (1); nanomaterials

Mathematics and Information

Statistical handling of data and statistical methods (3); cyber infrastructure (data preservation, processing, access) (9); theory and modeling of scaling (4)

Meteorology and Geology

Trace component behavior (2); control of pollutant plumes in groundwater (1); biogeochemical cycles (10); aerosol chemistry and clouds (2); chemical weather (1); chemistry-climate links (2); air pollution processes (1)

Biology

Life in extreme environments (2); scale-up of bench- to full-scale processes (2); application of microbiological biotransformations to industrial processes (2); conversion from batch to continuous processes (1); bioavailability (4); structure-activity relationships (8); new enzymes; bioinformatics (4); proteomics and genomics (4)

Problems and Issues

Professional societies and journals; interdisciplinary and multidisciplinary education (7); team approach; collaboration between development stages; completeness of life-cycle analysis (1); stovepiping of funding (2); breaking down administrative barriers (1); changing the scientific-academic culture (motivation) (7); reward structure; industry-academic collaboration; communication (language) (1); regulation and science (1); instrumentation development; theory and modeling; education of public and politicians (3); grand challenges must be defined and funded (10)

Green Group Report

The blue group reported its analysis of interfaces as a matrix in which the rows are a series of environmentally important chemistry and chemical engineering research areas, and the columns are the other disciplines that will participate in the research activity (Table G-2).

Blue Group Report

Biology (8)

Science (microbial community genomics; microbial in situ bioremediation; proteomics and metabolomics; PCR and revolutions in microbiology (6); engineering (bioprocessing, biotechnology, combining unit operations) (2)

Physics and Other Engineering (5)

Tools (new instrumentation; measurement systems and platforms; materials and processes for alternative, nonfossil fuels)

Atmospheric Science (5)

Modeling (chemical weather forecasting; air quality and human health; dispersion of chemicals; aerosol and cloud physics)

TABLE G-2 Matrix of Interfaces

	Mathematics	Computational Science	Hydrology and Geology	Atmospheric Science	Biology and Medicine
Advanced simulations	X	X			
Multiscale computing	X	X			
Dispersion of chemicals			X	X	
Chemical weather forecasting		X		X	
Cloud physics and aerosols					
Air quality and human health					X
Microbial community genomics					X
Proteomics					X
PCR microbiology					X
Climate questions		X		X	
Analysis of high-throughput datasets	X	X			
Instrumentation measurement systems					
Microbial remediation					X
Bioprocessing					X
Biotechnology					X
Chemical footprint of society					
Nonfossil energy					

Other Engineering	Physics	Ecology	Oceanography	Toxicology	Social Sciences and Economics
X	X				
X	X				
			X		
	X		X		
	X		X		
		X		X	
		X	X		
X					
X	X				
		X		X	X
X	X				X

Hydrology, Geology, Oceanography (5)

Dispersion of chemicals in groundwater and subsurface water (3); global cycling (2)

Ecology (4)

Climate questions (4); ecotoxicology

Social Sciences (3)

Chemical footprint of society (2); integrated assessment (1)

Computer Science (2)

Hardware: advanced simulations on high-performance computers; design of new computer architectures; software: analysis of high-throughput datasets (1)

Mathematics (2)

Multiscale computing (time and space)

Agriculture and Soil Science (11)

Closed-system agriculture (recycle all); precision farming (e.g., using GPC to improve farming); advances in pesticides, herbicide, and fertilizers; avoiding agricultural runoff; reducing agricultural footprint (largest impact on the environment)

Biology (8)

Biochemistry and biocatalysis; (green chemistry); genetic engineering; genomics and proteomics; epidemiology; effects of contaminants on humans; biomimetic processes

Earth Sciences (7)

Biogeochemical and Earth systems analysis; mechanisms and impact of natural contaminants; global cycles; C, N, etc.; subsurface detection and mapping; ocean ecology; climate; solid waste management; actinides in the environment

Engineering (5)

Process monitoring; sensors, spectroscopy, end users; process development (batch to continuous; throughput); material sciences; bringing chemistry to other engineering disciplines; systems and life-cycle analysis

Mathematics, Computational Science, Statistics (2)

Bioinformatics (handling huge databases); integration and mining of data sets (how to use the datasets); modeling

Space Sciences (2)

Remote sensing; recognition of rate of change on Earth; astrophysics (chemistry, biology)

Physics (1)

Moving technology out of hands of physicists; green chemistry for microelectronics; amorphous systems

All + Other (overarching)

Surface science; analytical techniques applied at the interfaces; integration with business model (economics); combinational chemistry at the interface with other disciplines; reduction of wastes

Grand Challenges

Integrated environmental education; exploiting natural remediation process; agricultural as a closed system; development of renewable energy sources; development of renewable chemical feedstocks economically; mimicking natural components

Black Group Report

Technical Solutions

Biochemists, molecular biologists, separation scientists, physicists, material scientists (bio-based approaches; physics and material approaches)

Risk Analysis

Medicine and health, social sciences, ecology, economics (health effects; ecological effects; social and political effects)

State of the Environment

Biologists, computer science, ecologists, meteorologists, instrumentation scientists, physicists (air quality; global warming; water quality; life-cycle analysis; biogeochemical cycles)

CHALLENGES

What are the environment-related grand challenges in the chemical sciences and engineering?

Red Group Report

Industrial Sustainability (1)

Ionic Liquids

Life cycle (1)

Remediation

Cost-effective methods

Contaminant Capture or Sequestration (3)

(Physical Chemistry)
CO_2; Radioactive waste

Lifecycle of Particulates (2)

Properties and effects on health and climate

Fuel Cells

H_2 Economy (2)

Environmental footprint

Biogeochemical Cycle

Role of chemical cycles on climate, greenhouse gases, clouds

Heterogeneous Characteristics of Environmental Materials (e.g., solids) (7)

Characterization of the Organic Composition of the Atmosphere

Sensing techniques; instrumentation

Environmental Change (Indicators) (2)

Traces in ice, trees, sediments; other indicators

Water (5)

Purification; industrial use; desalinization

Toxicity

Identify materials

Computational chemistry (1)

Instrumentation at Trace Levels (6)

Methods that will allow breakthroughs in air sampling

Catalysis by Design (3)

Instrumentation; computation; materials science

Function and Sensitivity of Nonculturable Organisms (1)

Photoinduced Processes

Macromolecular Science (3)

Biological molecules; catalysts; humic

Understanding Chemical Reactivity (7)

Complex processes in atmosphere; water; soil; all of the preceding

Study of Heterogeneous Processes (1)

Diversion of Funding of Core Sciences to Hot Initiatives

Separations (3)

Water purification; dilute solutions; air; bioproducts

Green Group Report

Global-Scale Chemistry (11)

Global biogeochemistry; flux measurements; prediction, fate, and transport of pollutants; speciation of all contaminants; chemistry and climate; carbon management; scaling in time and space; global-scale chemical monitoring

Waste Management (6)

Waste disposal technology; chemical; nuclear; mining; agriculture; animal residue; carbon sequestration; in situ remediation of contaminated media

Control and Understand Chemical Transformation (6)

Early life-cycle analysis; pollution prevention through alternative chemical processes; technologies for water conservation; controlled oxidation of organic molecules; end-to-end chemical production processes

Health effects (1)

Molecular basis for dose-response relationships; quantification of health risks for particulate matter

New Tools (9)

Better computational tools; acquiring relevant high-quality datasets; enhanced analytical capabilities; better models, theory, instrumentation; better analytical-theoretical tools to understand chemistry at interfaces

Blue Group Report

Fundamental Understanding

Understanding climate and chemical links (11)
  • Unraveling mechanisms: chemical reactions and cycles in the environment
  • Characterization of chemical processes in the atmosphere, biosphere, and lithosphere
Measurement of trace reactive species in atmosphere, ground- and surface water (7)
Sensor development for subsurface (e.g., pH, flow, microbial activity, pollutants, metals (6)
Better instruments for monitoring
  • Predictive chemical modeling of biological chemistry (4)
  • Long-term monitoring and tools, networks (1)

Chemical Approaches to Solutions

Predictive modeling for catalysts and custom-designed enzymes for bioremediation (10)
Chemical engineering pathways to environmental sustainability (5)
Conversion of solar energy into electrical and chemical energy (renewable) (4)

Transformation of complex agromaterials into advanced products (3)
New materials: designed for recycling (triggerable disassembly) (3)
Conversion of wastes to useful products (2)
Organic reductive chemistry for biorenewable feedstocks (2)
Remediation of mercury (1)
Mimicking natural processes
Using methane and hydrogen sulfide as feedstocks

Black Group Report

Enablers

Computing (hardware and software); genomic sciences; nanotechnologies; materials by design; self-assembly; instrumentation (in situ measurement of pollutants)

Energy

Global warming mitigation; carbon sequestration (6); economical solar energy (6); energy efficiency technologies (2)

Chemical-Organism Interactions

Molecular toxicology (5); fundamental understanding of microbes/metabolic pathways—"Perry's Handbook for Bugs" (4); models of exposure effects

Air and Water Issues

Understanding heterogeneous chemistry (4); the structure of natural organic matter (2); air: sources and characterization of toxics (1); speciation of toxic metals (1); smokestack emissions beyond SO_x and NO_x (1); chemical sources of toxicity of fine particulate matter (PM2.5) (1); environmentally persistent free radicals (1); understanding of greenhouse gases in the atmosphere (carbon cycles and sinks)

Remediation

Metabolic engineering—understanding pathways (3); economical metal sequestration methods (3); dioxins, MTBE (1); defluorination of fluoroorganics

Green Chemistry, Pollution Prevention

Catalysis by design (7); solvent-free processes (2); persistence of chemicals in the environment (1); alternative materials—structure-function relations (1); microreactors—just-in-time and local manufacturing (1); carbon source for chemicals

Fundamental Understanding

Techniques for assessing multisource risk (6); multimedia multiscale systems modeling (5); understanding natural environmental processes (3); control of chemical transformation (3); modeling of episodic events in the climate and environment (1); genetic and proteomic markers

INFRASTRUCTURE

What are the issues at the intersection of environmental studies and the chemical sciences for which there are structural challenges and opportunities—in teaching, research, equipment, codes and software, facilities, and personnel?

Red Group Report

What's Working Well

Some user facilities such as SSRL, EMSL computational facilities; ACS environmental chemistry option; green chemistry funding has raised profile (but may have taken from other environmental programs); multidisciplinary programs

What's Not Working Well?

EPA: lack of internal coordination and communication; deterioration of scientific facilities; coordination of environmental research effort—intra- and interagency; continuity of funding; user facilities—ease of access; principal investigator reward structure in national laboratories; lack of emphasis on environmental chemistry in NSF; initiatives—however valuable—have come at the expense of core sciences and engineering; lack of overall strategy—we are taking away from core to fund fads; disconnect between graduate research and undergraduate curriculum

Needs

User facility for geosciences (set of field sites); strengthen EPA R&D; an agency looking at environment over the long term; environmental equivalent of NIH (i.e., relationship to EPA as NIH is to FDA); educate people with broad-spectrum knowledge of the environment; curriculum reform to emphasize the environment; chemical programs to incorporate broad range of topics that are part of environmental chemistry; continued development of sensors and of instrumentation that is deployable in real time (for water); general support for instrument development (as in the medical field); interdisciplinary systems approach—resources and reward systems

Challenges

Inter- and intra-agency coordination; gulf between environmental sciences and others; market environmental chemistry as NIH does for health; environmental chemistry not well-regarded compared to "real" chemistry; atmospheric chemistry more accepted in academia than condensed-phase studies; if funding pot doesn't grow, allocation needs to be done smartly

Payoffs

Fundamental lead to prediction; our well-being is at risk and must be addressed; better use of resources; social and political stability; ability to predict and plan for environmental change on global scale (could prevent global upheaval)

A Final Word...

The environment should not be considered a fad!

Green Group Report

Student training in the environmental area (7)
Tenure process inhibits interdisciplinary collaboration (5)
Insufficient emphasis on classical measurements (5)
    Kinetics; thermochemical
Environmental chemistry is not considered a core subject in chemistry and chemical engineering departments—need accreditation (5)
    Coordination and interaction of similar programs across federal agencies (4)
Research needs (3)
    Aircraft, ships, space
Status of environmental chemistry in universities (e.g., few chemistry departments have major activities in the field) (2)
    Transitioning-leveraging: need for engineering expertise to develop instruments (2)
    Concepts from researchers; how to implement the design and construction
Need environmental science home in NSF for sustained funding of individual investigators (2)
    Heavy-metal actinide chemistry (expertise is being lost)
    Increased NIH funding has put a strain on national centers (computing, spectroscopy, etc.) for environmental projects

Blue Group Report

There is no single federal agency that is responsible for funding environmental science. EPA is an environmental regulatory agency! *There is a pressing need for one centralized government agency for funding fundamental science with respect to the environment.*

The NSF Science and Technology Centers and Environmental Molecular Science Institutes and Collaborative Research in Environmental Molecular Science Grants Program are small steps toward improved funding in the environmental science research area.

Sustained long-term support for individual investigators in the environmental chemistry/science area is needed (applies also to technicians and support staff for research groups.)

Industry is a failure in environmental science R&D area and in providing attractive jobs for environmental chemistry students! (Not all in blue group agree with this statement).

Mechanisms for attracting more young people to environmental science and the basic sciences are needed. Good jobs are necessary to achieve this need.

Cross-training of students among interdisciplinary areas is lagging far behind need.

Basic science and math training is absolutely essential for students in environmental science.

There is only one national facility dedicated to environmental science research (EMSL at PNNL)

Faculty in basic science departments who focus on environmental science research often have difficulties at tenure decision time because of a general lack of respect for environmental science research by other scientists.

Physical chemistry is not thriving in modern research universities, yet it provides the basic underpinnings of environmental chemistry.

Research agendas should be driven in part by societal needs (e.g., human health), but basic, curiosity-driven research funding must be protected and allowed to thrive.

More scientific expertise is needed in policy making.

Black Group Report

Agency Funding and Missions

Continuing mismatch between core competencies and problem-driven research (programs, time frames, etc.); chemistry and chemical engineering subdivisions not coherent with environmental problems; funding sources need to be tied to problem horizons; need to avoid discontinuities for funding specific areas;

need more sustained funding for instruments and instrumentation; mission alignment—pressure on regulators to declare problems solved; need for continuity for long-term monitoring; need appropriate mixture of individual investigators and centers to meet various challenges

Chemistry Chemical Engineering Education

Chemistry and chemical engineering subdivisions not coherent with environmental problems; environmental aspects of chemistry and chemical engineering not always treated or sold well; barriers to cross-disciplinary work; need more use of environmental examples at undergraduate level; greater need to align multidisciplinary programs and expectations of graduate students

Tools

Structure-activity relationship models are too empirical

Things That Are Working Well

Increases in joint agency funding (clarity of purpose); utilization of SBIR; funding of large field studies; centers—some successes

H

List of Acronyms

ACS	American Chemical Society
3G	1,3-propanediol
BHC	brominated hydrocarbon
CFC	chlorofluorocarbon
CHC	chlorinated hydrocarbon
CIRPAS	Center for Interdisciplinary Remotely-Piloted Aircraft Studies
CMAQ	community modeling of air quality
DDT	dichlorodiphenyltrichloroethane
DEA	Drug Enforcement Agency
DOD	Department of Defense
DOE	Department of Energy
DOM	dissolved organic matter
DSIDA	disodium iminodiacetic acid
EDC	endocrine disrupting chemical
EMSL	Environmental Molecular Sciences Laboratory
EPA	Environmental Protection Agency
EPR	electron paramagnetic resonance
FDA	Food and Drug Administration
FOMA	1,1-dihydroperfluorooctyl methacrylate

GC-MS	gas chromatography–mass spectrometry
GOME	Global Ozone Monitoring Experiment
GPC	gas-to-particle conversion
GPS	Global Positioning System
HCCI	homogeneous-charge compression ignition
HCFC	hydrochlorofluorocarbon
HFC	hydrofluorocarbon
INDOEX	Indian Ocean Experiment
IPCC	Intergovernmental Panel on Climate Change
ITCZ	Intertropical Convergence Zone
MTBE	methyl *tertiary*-butyl ether
NASA	National Aeronautics and Space Administration
NCAR	National Center for Atmospheric Research
NIH	National Institutes of Health
NMR	nuclear magnetic resonance
NOAA	National Oceanic and Atmospheric Administration
NREL	National Renewable Energy Laboratory
NSF	National Science Foundation
ORD	Office of Research and Development
PAG	photoacid generator
PAH	polycyclic aromatic hydrocarbon
PBDE	polybrominated diphenyl ether
PCB	polychlorinated biphenyl
PCDD/F	polychlorinated dibenzodioxins and furans
PCR	polymerase chain reaction
PM	particulate matter
PM2.5	fine particulate matter (smaller than 2.5 microns in diameter)
PNNL	Pacific Northwest National Laboratory
ppm	parts per million
ppt	parts per trillion
psi	pounds per square inch
SBIR	Small Business Innovation Research Program
SCR	selective catalytic reduction
SSRL	Stanford Synchrotron Radiation Laboratory
STAR	Science to Achieve Results (an EPA grant program)
SVA/lb	shareholder value added per pound

THPMA   2-tetrahydropyranyl methacrylate

USGS   U.S. Geological Survey

VOC   volatile organic compound

XAS   x-ray absorption spectroscopy